Nelson Advanced Science

Structure, Bonding and Main Group Chemistry

revised edition

Rod Beavon • Alan Jarvis

Endorsed by

First published in 2000 by:
Nelson Thornes Ltd
Delta Place
27 Bath Road
CHELTENHAM
GL53 7TH
United Kingdom

This edition published in 2003

05 06 07 / 10 9 8 7 6 5 4 3 2

A catalogue record for this book is available from the British Library

ISBN 0 7487 7655 9

Illustrations by Hardlines and Wearset
Page make-up by Hardlines and Wearset

Printed and Bound in Croatia by Zrinski

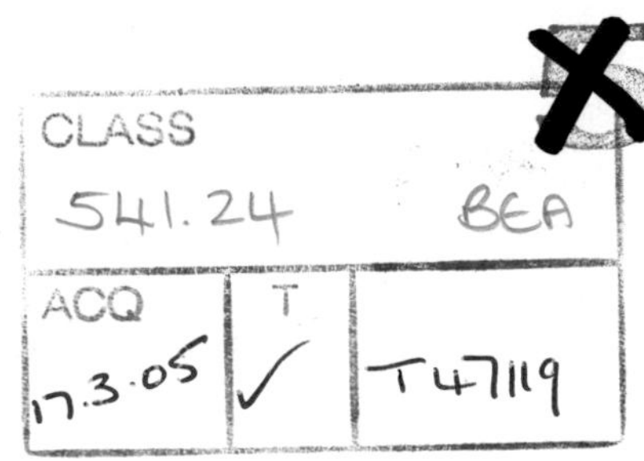

Contents

Introduction

Introduction

This series has been written by Examiners and others involved directly with the development of the Edexcel Advanced Subsidiary (AS) and Advanced (A) GCE Chemistry specifications.

Structure, Bonding and Main Group Chemistry is one of four books in the Nelson Advanced Science (NAS) series developed by updating and reorganising the material from the Nelson Advanced Modular Science (AMS) books to align with the requirements of the Edexcel specifications from September 2000. The books will also be useful for other AS and Advanced courses.

Structure, Bonding and Main Group Chemistry provides coverage of Unit 1 of the Edexcel specification. It deals with fundamental principles applicable to the whole of chemistry, together with introductory inorganic chemistry of some main-group elements. Atomic structure is developed so that the subsequent ideas of structure, bonding and the resulting properties of substances can be correctly appreciated. Chemistry is a quantitative subject; one chapter is concerned with chemical calculations, and offers many ideas for making what is sometimes thought to be a difficult area of chemistry easier. Then a consideration of oxidation and reduction prepares the ground for much other factual chemistry. A short overview of one period in the Periodic Table is supplemented by a more detailed consideration of the chemistry of the alkali metals, the alkaline earth metals and the halogens. This is related to the general principles elucidated earlier.

Other resources in this series

NAS *Teachers' Guide for AS and A2 Chemistry* provides advice on safety and risk assessment, suggestions for practical work, key skills opportunities and answers to all the practice and assessment questions provided in *Structure, Bonding and Main Group Chemistry; Organic Chemistry, Energetics, Kinetics and Equilibrium; Periodicity, Quantitative Equilibria and Functional Group Chemistry;* and *Transition Metals, Quantitative Kinetics and Applied Organic Chemistry.*

NAS *Make the Grade in AS and A Chemistry* is a Revision Guide for students. It has been written to be used in conjunction with the other books in this series. It helps students to develop strategies for learning and revision, to check their knowledge and understanding and to practise the skills required for tackling assessment questions.

Features used in this book

The Nelson Advanced Science series contains particular features to help you understand and learn the information provided in the books, and to help you to apply the information to your coursework.

These are the features that you will find in the Nelson Advanced Science Chemistry series:

INTRODUCTION

Text encapsulates the necessary study for the Unit. Important terms are indicated in **bold**.

5 Oxidation/reduction: an introduction

DEFINITION

Oxidation is electron loss.
Reduction is electron gain.

MNEMONIC

OIL RIG:

oxidation **is l**oss

reduction **is g**ain

Introduction

Oxidation and reduction are found with all but four elements in the Periodic Table, not just with the transition metals, although they show these reactions to such an extent that they could be accused of self-indulgence.

When magnesium reacts with oxygen (Figure 5.1)

$$2Mg(s) + O_2(g) \rightarrow 2MgO(s)$$

the product contains Mg^{2+} and O^{2-} ions. Reaction with oxygen is pretty clearly oxidation. The reaction of magnesium with chlorine

$$Mg(s) + Cl_2(g) \rightarrow MgCl_2(s)$$

gives a compound with Mg^{2+} and Cl^- ions. In both cases the magnesium atom has lost electrons, so as far as the magnesium is concerned the reactions are the same. This idea is generalised into the definition of oxidation as loss of electrons. Reduction is therefore the gain of electrons. Since electrons don't vanish from the universe, oxidation and reduction occur together in **redox** reactions.

Figure 5.1 The use of magnesium flares in photography being demonstrated at an early meeting of the British Association in Birmingham (1865).

Oxidation numbers

For simple monatomic ions such as Fe^{2+} it's easy to see when they are oxidised (to Fe^{3+}) or reduced (to Fe). For ions such as NO_3^- or SO_3^{2-} which also undergo oxidation and reduction it is not always so easy to see what is happening in terms of electrons. To assist this, the idea of **oxidation number** or **oxidation state** is used. The two terms are usually used interchangeably, so that an atom may have a particular oxidation number or be in a particular oxidation state.

Definition boxes in the margin highlight some important terms.

Background boxes give further information to help you understand the topic. This information puts the topic into a broader context, but is not strictly part of the Edexcel specification.

DEFINITION

First ionisation energy: the amount of energy required per mole to remove an electron from each atom in the gas phase to form a singly positive ion, that is

$$M(g) \rightarrow M^+(g) + e^-$$

Second ionisation energy: the energy per mole for the process

$$M^+(g) \rightarrow M^{2+}(g) + e^-$$

and so on for successive ionisation energies.

Questions in the margin will give you the opportunity to apply the information presented in the adjacent text.

The **empirical formula** shows the ratio of atoms present in their lowest terms, i.e. smallest numbers. Any compound having one hydrogen atom for every carbon atom will have the empirical formula CH; calculation of the **molecular formula** will need extra information, since ethyne, C_2H_2, cyclobutadiene, C_4H_4, and benzene, C_6H_6, all have CH as their empirical formula. Empirical formulae are initially found by analysing a substance for each element as a percentage by mass.

QUESTION

Find the empirical formula of the compound containing C 22.02%, H 4.59%, Br 73.39% by mass.

Practice questions are provided at the end of each chapter. These will give you the opportunity to check your knowledge and understanding of topics from within the chapter.

Assessment questions are found at the end of the book. These are typical of the assessment questions for Advanced Subsidiary that you will encounter in your Unit Tests (exams) and they will help you to develop the skills required for these types of questions.

Questions

1. Plot on a (small) graph the first ionisation energies of the elements from sodium to argon, and account for the shape obtained.
2. Use data from a data book to plot a graph of atomic radius vs atomic number for the elements of Periods 2 and 3 (Li to Ar). Account for the difference in the graphs between Groups 2 and 3.
3. Sketch the structures of:
   (a) the giant covalent lattice of silicon
   (b) the molecule P_4
   (c) the molecule S_8.
4. Silicon has no compounds in which the silicon atom forms double bonds with other elements. Phosphorus, by contrast, does form double bonds with other elements. Suggest why silicon and phosphorus are different in this respect.

Acknowledgements

The authors and publisher would like to thank Geoff Barraclough for his work as Series Editor for the original series of four NAMS books, from which the new suite of NAS books was developed.
I want to thank the following most sincerely for their invaluable help and advice: at Westminister, Peter Hughes, Derek Stebbens, Martin Robinson, Gilly French, Damian Riddle and Nick Hinze; at Edexcel, Ray Vincent; and of course Alan Jarvis, for his friendship and advice during the years we worked together. My students have taught me more than almost anyone else – to them I owe a particular debt. Thanks too to my late wife Doreen, whose support was so welcome during the production of the first two editions of this book.

The authors and publishers would like to thank the following for permission to reproduce copyright material:

Photographs
Science Photolibrary: cover Ken Eward, 1.6, 1.9, 6.4 Department of Physics/Imperial College, 1.10 James Holmes/Oxford Centre for Molecular Sciences, 3.1, 3.18a, 3.18b Alfred Pasieka, 3.5 Adam Hart-Davis, 3.7 Thomas Hollyman, 6.3a David Taylor, 7.1 Jerry Mason, 7.2 Andrew McClenaghan 7.3 Claude Nuridsany & Marie Perennou, 1.1, 1.3, 1.5, 3.19, 6.3b, 6.3c, 6.3d, 6.3e, 6.3f

Science & Society Picture Library: page vi, 5.1;

Examination questions
Edexcel

About the Authors

Rod Beavon is Chief Examiner in Chemistry for Edexcel and Head of Science at Westminster School, London.

The late **Alan Jarvis** was former Head of Chemistry at Stoke-on-Trent Sixth Form and was Chief Examiner in Chemistry for Edexcel.

1 Atomic structure

Learning science

As you start this book, and probably the whole course, you may wonder what there is to be learned on the structure of atoms that is new. Maybe you feel you have already done it all. During the next months and years you will be learning more chemistry, but when you study a science, unlike say learning a language, the ideas which you already have about the world are modified and replaced by those which are presently accepted by scientists generally.

That is what we mean by being 'right'; teachers teach, and examiners mark scripts, on the basis of what is *currently accepted* as being true. Science itself is continually changing at its frontiers, and those who work with totally new material cannot appeal to some other authority for a judgement on what is right or not. All they can do is more work, keeping going, until it may become clear that others agree with their results, that their experiments are repeatable, and eventually their work is absorbed into the body of scientific knowledge. Learning science is also a replacement activity, and because some ideas are very strange compared with our everyday experience, it is quite hard, and requires a lot of work right from the beginning. Difficult ideas tend to be introduced in stages, and modified as you go through them. Some people don't like this, thinking that their key stage 4 picture of the atom is in some way a lie, and that the A level view will be too. But the pictures that you learn are approximations, which you can build on if you need to, or hopefully will want to, as your general scientific and mathematical abilities increase and mature. In many ways your progress through chemical ideas will parallel the way that these ideas have developed historically; your learning of science is similar to the way in which society has learned science – though I hope much faster!

So that we can deal with the real world of experiment, we have to disregard the imperfections which we can ascribe to 'experimental error', but we also have to learn how to know when odd or inconsistent results are due to some underlying and important feature rather than error. A rather unnerving aspect of chemistry is that apparently small differences in properties, say in energy changes or electron structures, can affect significantly the chemistry which is seen in the test tube.

Although general rules of chemical behaviour are put forward, there are usually exceptions. Thus chemistry needs a broad approach at A level, where many of the exceptions are not considered; if they were, the syllabus would be immense. For this reason, subtle effects are sometimes ignored, but this is done in such a way that if you go on to do more chemistry, you will not have to unlearn anything.

Figure 1.1 John Dalton, the originator of modern atomic theory.

Before we start the atomic structure unit, here are some hints. Chemistry is often seen by non-chemists to be a manipulation of formulae on paper, as mysterious, alchemical almost. If you still think this, do the following:

- Imagine yourself as being of atomic or molecular size. This is not easy, but essential to an understanding of what happens to particles as they collide and react.
- Picture in your mind the appearance of the materials that occur in the equations that you write, how they are reacting, the apparatus that you might use. If you don't know some of these things, find out. It is vital that the equations are related to real events; chemistry happens in glassware, not on the pages of a textbook.
- Consider how chemistry affects everyday life; be aware of the materials you use and why, of the impact of chemistry all around you.

BACKGROUND

Atomic structure

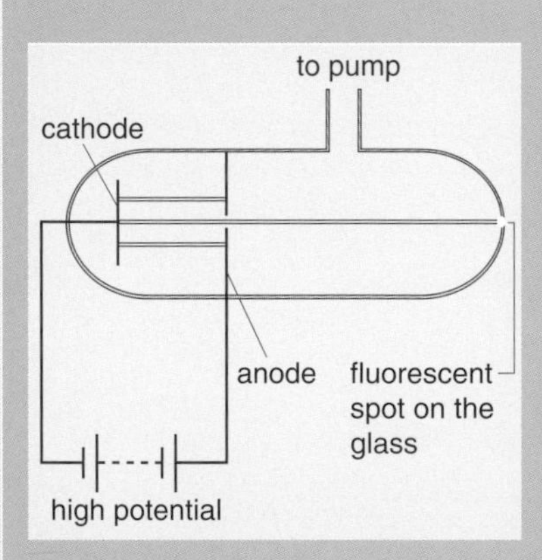

Figure 1.2 The production of cathode rays.

As you work, remember that this material took hundreds of people some 130 years to discover; however you will find out by the end of Chapter 3 why you can't pull many everyday objects apart very easily, and yet don't permanently bond to the chair on which you are probably now sitting.

The (much shortened) history of the discovery of the atom

Democritus in 585 BC used the word ατομος, atom, for the first time. If you read *On the Nature of the Universe* by Lucretius, many of the passages have a familiar ring to them. But the modern experimental determination of atomic structure, as distinct from the speculative writings of the Greeks, started in 1808 with John Dalton (Figure 1.1). His notion that all substances were made of extremely small, indivisible particles was highly controversial at the time, and there were many highly respected nineteenth century chemists who did not believe Dalton's hypothesis, even to the ends of their lives.

Figure 1.3 Pierre and Marie Curie, discoverers of polonium and radium.

The investigation of atomic structure got nowhere, mainly because of the technological problems. How should the experiments be done? With what tools could the atom be investigated? If you look back over the history of the discovery of the elements, you find that there are spurts of activity that follow a technological advance. Thus the invention of the battery by Volta led rapidly to electrolysis and to the isolation of the alkali and alkaline earth metals. Developments in high-vacuum technology enabled J. Thomson in 1897 to find the ratio of the charge to mass of the cathode rays which were found when a high potential was placed across a gas at low pressure (Figure 1.2), and which had been known about for some time. The particle was shown to be common to all the gases used, and was called the **electron**.

Similar experiments were performed with positively charged particles which were generated in the same experiment by reversing the polarity of the electrodes. The masses of these particles were dependent on the gas in the tube, and Thomson found that the lightest of them came from hydrogen. He called this particle the **proton**, and speculated that an atom consisted of a sphere of protons with electrons embedded in the sphere. This famous 'plum-pudding' model of the atom could not be tested until the tools were available; these came from the Curies.

The discovery of radium provides the tool to show atomic structure

In 1896 A. H. Becquerel had discovered radioactivity. He had been investigating fluorescence in salts; uranium salts fluoresce strongly, and he had left some in a drawer with a key and a photographic plate wrapped in light-proof wrapping. There was an image of the key on the plate when it was developed, due to the radiation having penetrated through to the plate and so forming the image. In 1902 the Curies had taken on the enormous task of isolating the radioactive elements, and produced the first samples of polonium and of radium. Becquerel received the 1903 Nobel Physics prize jointly with Pierre and Marie Curie (Figure 1.3). Marie died from leukaemia, caused by radiation, in 1934. We now know that radioactivity, X-rays, and many compounds are damaging or fatal, many people having died in the course of their scientific work.

The 'gold foil' experiment

The availability of radium, which emits alpha (α) particles, or helium nuclei, was crucial to the work of Rutherford, Geiger and Marsden who, in the years 1909 to 1911, fired α-particles at metal foils to see how they would be deflected or absorbed (Figure 1.4). If Thomson's model of the atom was correct, the α-particles would have mostly been absorbed, and those which did penetrate the foil would have been deflected only very slightly. This was Rutherford's expectation, since Thomson's atom was very solid. But the majority of the α-particles passed straight through, and a very few bounced back. This so startled Rutherford that he made his now famous assertion that he would have been less surprised if a naval shell had been fired at a sheet of tissue paper and had bounced off. This experiment made possible not only the modern view of the atom as being essentially mostly empty, but also the measurement of the diameter and charge of its nucleus. Although it is usually called the 'gold foil' experiment, the nuclear size was first estimated for copper and aluminium, in both cases, as being about 10^{-14} m.

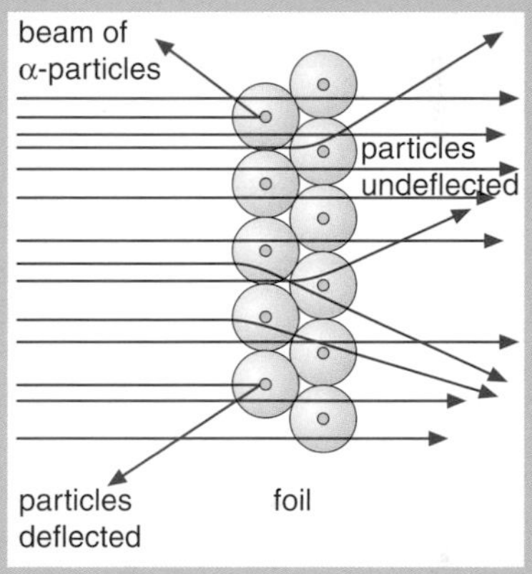

Figure 1.4 The 'gold foil' experiment elucidated the structure of the atom.

Investigations in the early 1880s by J. Balmer, a Swiss schoolteacher, on an apparently unrelated matter, enabled Niels Bohr (Figure 1.5) and his colleagues, some 25 years later, to suggest the way in which electrons are distributed in the atom. Balmer had shown that a gas excited electrically (such as in the sodium street lights with which you are doubtless familiar) gave out light which, when passed through a prism, did not show continuous bands of colour as the rainbow does, but rather a series of bright lines on a dark background. In the case of hydrogen these lines are quite widely spaced (Figures 1.6 and 1.7). Balmer was able to derive an equation for the frequencies of these lines, but he could not discover the meaning of his results since this required contributions from work which was not carried out until 1900, by Planck.

QUESTION

What sort of force causes the scattering of α-particles by nuclei?

Bohr's idea (Figure 1.5) was that the lines came about as excited atoms lost their excess energy as light, by the electrons falling from high energy levels in the atom to lower ones. The energy levels of the electrons were well-defined or

QUESTION

Which observation suggests that the nucleus is massive?

Figure 1.5 Niels Bohr, originator of the shell model of the atom.

Figure 1.6 The spectrum of excited hydrogen gas helped Niels Bohr determine how electrons are arranged in electron shells.

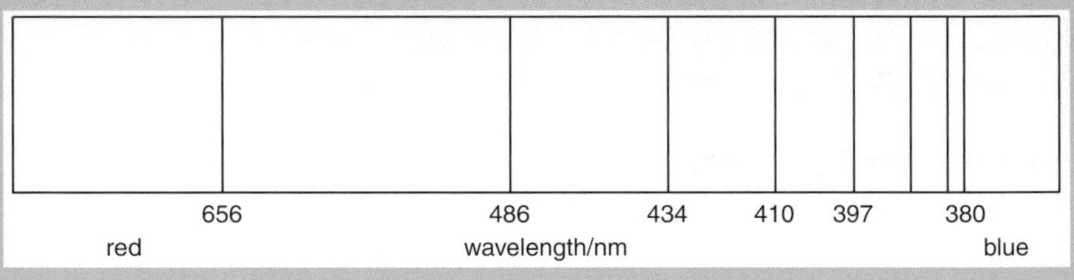

Figure 1.7 The Hydrogen spectrum, or Balmer series.

QUESTION

Spectra of atoms with two electrons or more are complex. Species having one electron, such as He^+, give simple spectra like that of hydrogen, but the lines have frequencies much higher than those of hydrogen. Can you suggest why?

quantised, and so, therefore, were the transitions between these levels, and this was shown by light (of a single frequency), and energy, being emitted for each transition. Bohr was the originator of the shell model of the atom, a sort of miniature solar system, which will be familiar to you. Bohr did not come to this conclusion in isolation, of course; he was able to draw on many other experiments.

The transitions which give rise to the visible spectrum of hydrogen, which is called the *Balmer series* of lines, are shown in Figure 1.8, as well as other series seen in other parts of the spectrum. Further evidence from work on the shell model comes from ionisation energies (discussed later in this chapter).

QUESTION

The energy of a given transition in the hydrogen spectrum is given by $E = hf$, where h = Planck's constant and f is the frequency of the radiation. What is the energy per atom for the transition giving the red line in the hydrogen spectrum? The frequency of this line is 4.573×10^{14} Hz, and $h = 6.262 \times 10^{-34}$ J s.

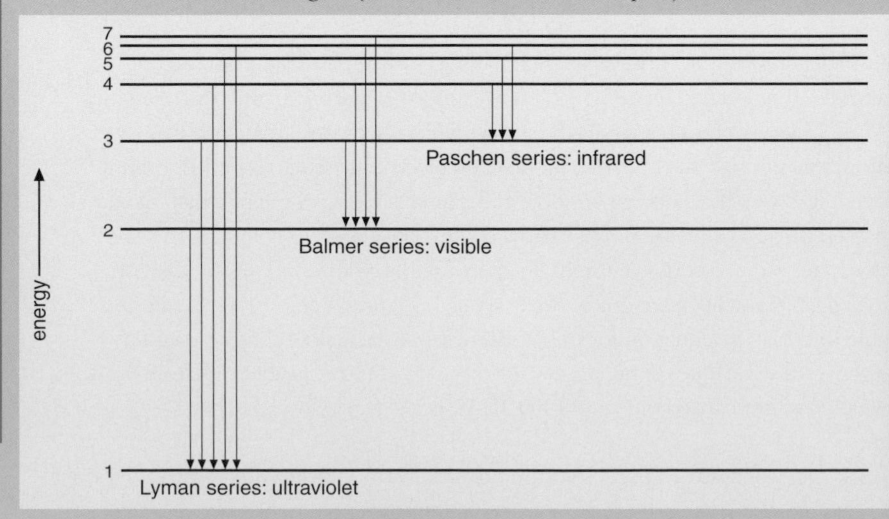

Figure 1.8 Electron transitions in hydrogen.

Bohr's theory was interesting, partly because he knew when he published it that it had serious imperfections. The main problem was that an orbiting electron is an accelerating charge, and accelerating charges radiate energy. If the electron in an atom did this, it would spiral in to the nucleus, and we would have no atoms. This problem was resolved by Schrödinger (Figure 1.9) and Heisenberg in 1928 when they published their theory of quantum mechanics, considered briefly in the section on atomic orbitals in this chapter.

The discovery of the neutron

The last piece of the atomic jigsaw (at least as far as chemistry is concerned) was placed when the neutron was discovered by Chadwick in 1932. It had been known for decades that atoms were more massive than the number of protons in them would allow; that difference in mass was accounted for by the neutrons in the nucleus.

That is not the end of the story of the atom; what is the nucleus like? That is really a question for physics, and is being actively pursued now. This belongs elsewhere; we must now look at the chemical atom in more detail.

Figure 1.9 Erwin Schrödinger, originator of the wave-mechanics model of the atom.

Particles and structure

The particles

The atomic model consists of protons and neutrons in the nucleus, and electrons in shells around the nucleus. The principal properties of these particles are given in Table 1.1.

Table 1.1 *Properties of nuclear particles (C = coulomb)*

Particle	Mass		Charge	
	Actual/kg	Relative to proton	Actual/C	Relative to proton
proton, p	1.6726×10^{-27}	1	$+1.6 \times 10^{-19}$	+1
neutron, n	1.6750×10^{-27}	just over 1	0	0
electron, e	9.1095×10^{-31}	1/1836	-1.6×10^{-19}	−1

You should have an idea of the relative masses and charges of these particles, but you need not learn their actual values.

The nucleus

The protons and neutrons constitute the nucleus. As the number of protons increases, the number of neutrons increases relatively faster, so small atoms have proton and neutron numbers which are comparable, whereas large atoms have many more neutrons than protons. These neutrons act as a sort of diluent, reducing the repulsive forces between the positive protons. The ratio of protons to neutrons is fairly critical, and any departure from the optimum range will lead to nuclear instability and thus radioactivity.

The **atomic number**, Z, of an element is the number of protons in the nucleus of its atom, and it defines the element. The element with six protons in its nucleus must be carbon. The **mass number**, A, is the number of protons plus the number of neutrons; it is close to, but not identical with, the relative atomic mass, and is always a whole number since you cannot have part of a proton or part of a neutron.

DEFINITION

The **atomic number *Z*** (proton number) of an element is the number of protons in the nucleus of its atom.
The **mass number *A*** of the atom is the number of the protons plus the number of neutrons in the nucleus.

Symbolically, the element E would be represented $^{A}_{Z}E$. Thus for carbon, having in its most common isotope six protons and six neutrons, we write $^{12}_{6}C$.

The nucleus constitutes most of the mass of an atom. Although the mass of the electrons is not zero, for the majority of practical purposes their mass is ignored.

Isotopes

All atoms have isotopes. The word comes from the Greek for the 'same place', because all the isotopes of a given atom occupy the same place in the Periodic Table defined by the atomic number. Isotopes arise because of changes in the neutron number. This affects the mass of the atom, but it does not affect the electron number or structure, and so does not affect the chemistry of the element. (With isotopes which are significantly different in mass, in practice those of hydrogen, the *rates* of reaction may vary with the different isotopes, but not the *nature* of the reaction.) So, isotopes are atoms with the same atomic number but different mass numbers. Some examples of isotopes are mentioned later in this chapter.

DEFINITION

Isotopes are atoms having the same number of protons, but a different number of neutrons. They have the same atomic number but different mass numbers.

The word *isotope* does not imply radioactivity; carbon has three naturally occurring isotopes, only one of which is radioactive, and four artificial ones which are all radioactive. A nuclear species of a given mass number and atomic number is often called a **nuclide**; radioactive ones are **radionuclides**.

Relative atomic mass

Calculations in chemistry do not need the actual masses of atoms to be known, since they usually involve the relative proportions of the atoms concerned. The **relative atomic mass** is used where the unit of mass is defined as one-twelfth the mass of the carbon-12 isotope, whose mass is defined as 12 units exactly. The **relative atomic mass**, A_r, of an element is defined as the average mass of an atom of the element divided by the mass of 1/12 of the mass of an atom of ^{12}C.

DEFINITION

The **relative atomic mass** is the average mass of an atom of the element divided by the mass of 1/12 of the mass of an atom of ^{12}C.

It is important to be clear, when relative atomic masses are being used, whether the value is for a single pure isotope or for the naturally occurring mixture of isotopes of an element. The relative atomic mass of the isotope ^{12}C is 12.000 by definition; natural carbon contains some ^{13}C and ^{14}C as well, so that the relative atomic mass of natural carbon is 12.011. A greater deviation from whole numbers is seen with chlorine: 75% (or as a fraction, 0.75) is present as ^{35}Cl, and 25% (0.25) as ^{37}Cl. Natural chlorine, therefore, behaves as an element with a relative atomic mass of

$$(0.75 \times 35) + (0.25 \times 37) = 35.5$$

No chlorine atoms contain half a particle, however.

DEFINITION

For any element, E,

$$\text{relative atomic mass} = \frac{\text{mass of one atom of E}}{\text{1/12 mass of one atom of carbon-12}}$$

The relative molecular mass for a compound (sometimes called the formula mass, especially for ionic substances which have no molecules) is the sum of all the relative atomic masses of the elements which make up the compound's formula. Usually the use of masses to three significant figures is adequate.

The mass spectrometer

The mass spectrometer (Figure 1.10) is a device for measuring the masses of positive ions. The mass spectrograph, its forerunner, was invented by F.W. Aston in 1919, for which he received the 1922 Nobel Prize for Chemistry.

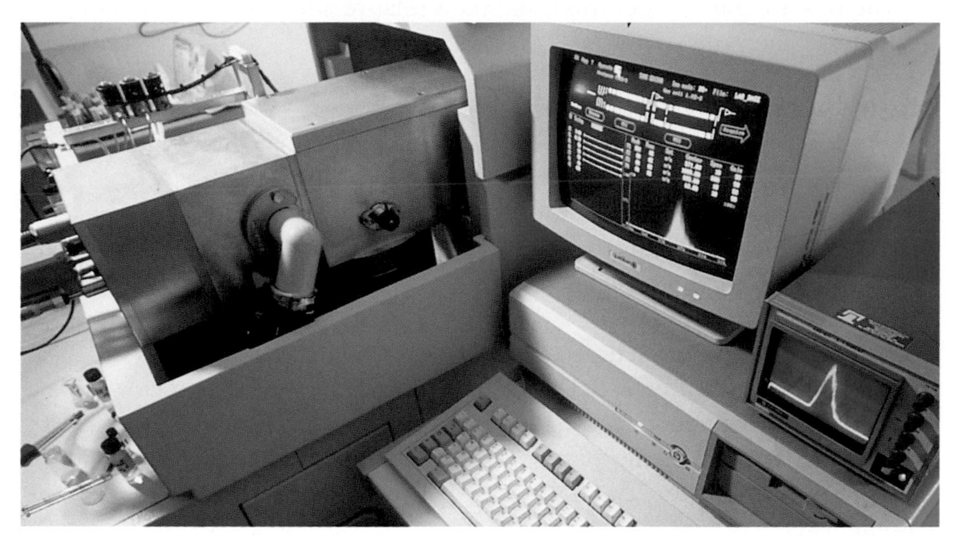

Figure 1.10 A mass spectrometer measures the mass of positive ions, thus helping to determine the structure of compounds.

Mass spectrometry makes it possible to determine the structure of compounds by studying the masses of the ions produced by a molecule and the ways in which the molecule fragments or rearranges in the machine.

The essential principles of the mass spectrometer (Figure 1.11) are as follows.

1. Electrons are emitted from the filament, accelerated, and used to bombard the gaseous sample which is at very low pressure.
2. Sample molecules have electrons knocked off them by the bombarding electrons, forming positive ions. The molecule can also fragment and possibly rearrange giving different positive ions.
3. The positive ions are accelerated by an electric field.

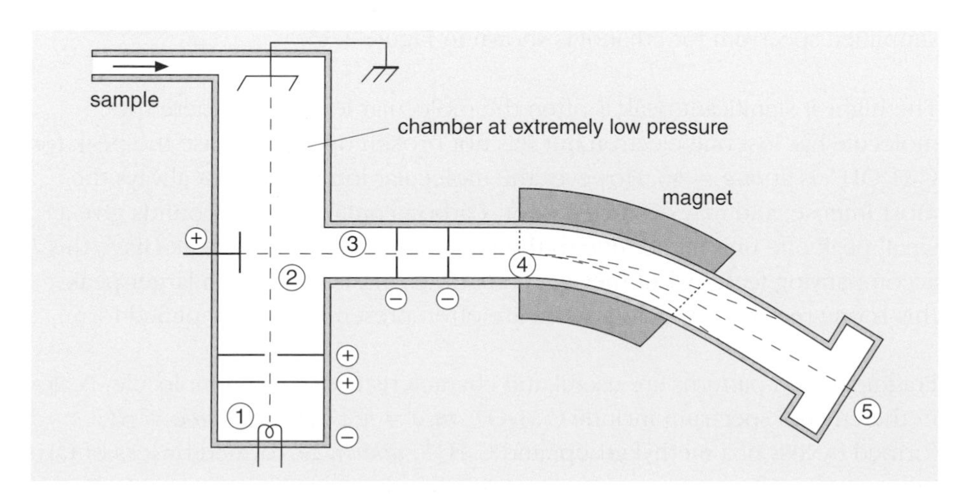

Figure 1.11 Principles of the mass spectrometer.

QUESTION

No element in the Periodic Table has a relative atomic mass that is a whole number. Why not?

QUESTION

At one time chemists used relative atomic masses based on oxygen = 16, whereas physicists used them based on hydrogen = 1. Why did the two scales not agree?

QUESTION

The electrons are accelerated in the mass spectrometer and pass through holes in the anodes. Why, once the electron has passed the final anode, does it not decelerate and come back to the anode?

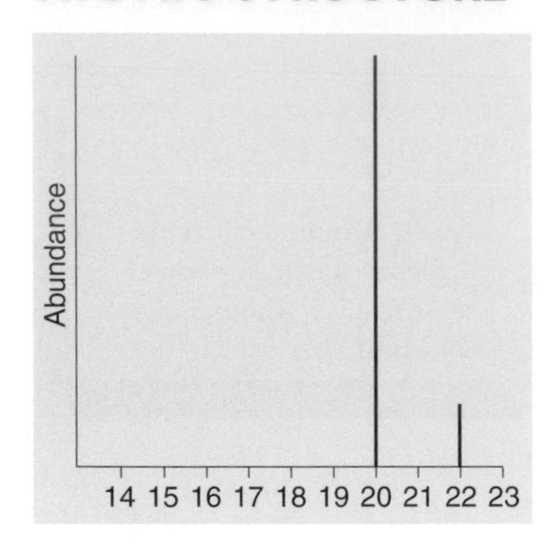

Figure 1.12 The mass spectrum of neon.

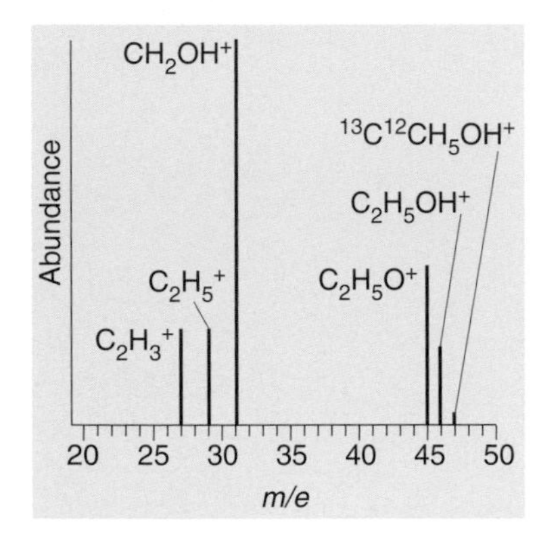

Figure 1.13 The mass spectrum of ethanol.

4 The positive ions are deflected by a magnetic field in a circular path whose radius depends on their mass/charge ratio and the strength of the field. The machine sweeps over the chosen mass range by altering the magnetic field, and hence the ions which reach the detector are separated according to their mass.

5 The positive ions are detected, and their relative amounts calculated by the machine.

Mass spectra

The mass spectrum is usually converted to a bar graph that shows the relative abundance of the various fragments detected. Because of the variety of fragmentation possible in larger molecules, not all the peaks are interpreted. Practical machines use additional features to ensure that all ions entering the magnetic field have the same energy.

The use of mass spectrometry enabled the detection of isotopes, and historically, neon was the first element to be so investigated. The relative atomic mass of naturally-occurring elements is not a whole number because it is a weighted mean, and the mass spectrum of neon (Figure 1.12) shows 90.9% of $^{20}Ne^+$, 0.26% of $^{21}Ne^+$, and 8.8% of $^{22}Ne^+$, so the relative atomic mass of the natural material is

$$(0.909 \times 20) + (0.0026 \times 21) + (0.088 \times 22) = 20.17$$

The horizontal axis is strictly mass/charge, m/e.

The mass spectrum of chlorine shows peaks due to Cl^+ and Cl_2^+; one line each from isotopes of masses 35 and 37, and three, from all pairings of these at masses 70, 72, and 74. Knowing the peak heights enables the abundance of each of the isotopes to be calculated. Mass spectrometry is very accurate; masses are generally quoted as whole numbers, but high-resolution machines can find masses to eight or ten decimal places. The production of singly-positive ions is assumed; the chances of removing two electrons by bombardment to form 2+ ions are very small.

The mass spectrometry of compounds is more complex, but informative; a simplified spectrum for ethanol is shown in Figure 1.13.

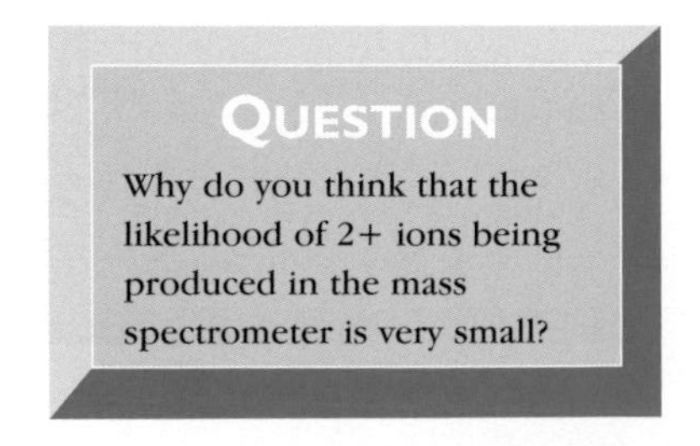

The highest significant peak is often the molecular ion peak, where the molecule has lost one electron but has not broken up. In this case the peak for $C_2H_5OH^+$ is at $m/e = 46$. However the molecular ion peak is not always the most intense, and may even be absent. Carbon-containing compounds give a small peak one unit higher due to the presence of ^{13}C. All the peaks have this accompanying feature, but they are sometimes masked by much larger peaks; this is one reason why mass spectra are often presented in a simplified form.

Fragmentation patterns are useful and characteristic of a given molecule. Peaks in the ethanol spectrum include $C_2H_5O^+$, $m/e = 45$; CH_2OH^+, $m/e = 31$, formed by loss of a methyl group; and $C_2H_5^+$, $m/e = 29$, formed by loss of OH.

Mass spectrometry is used, together with complementary evidence from other techniques such as infrared (See *Transition metals, Quantitative Kinetics and Applied Organic Chemistry*, Thomas Nelson and Sons Ltd, 2001), ultraviolet and nuclear magnetic resonance spectroscopy, to find the structure of compounds. High-resolution machines can distinguish between ions of different composition but which have the same mass on an integer scale.

The mass spectrometer is often coupled with gas–liquid chromatography, especially in forensic work, as a very powerful analytical technique for complex mixtures.

Samples tested for drugs, for example, can be introduced into the gas–liquid chromatograph and the various constituents separated. If the output gases are then led to a mass spectrometer, evidence from fragmentation patterns can be used to identify materials present in very small amounts. This, and the huge variety of compounds possibly present, makes mass spectrometry very useful, since there is a large database of patterns from known substances that can be used for comparison.

> **DEFINITION**
>
> **First ionisation energy:** the amount of energy required per mole to remove an electron from each atom in the gas phase to form a singly positive ion, that is
>
> $$M(g) \rightarrow M^+(g) + e^-$$
>
> **Second ionisation energy:** the energy required per mole for the process
>
> $$M^+(g) \rightarrow M^{2+}(g) + e^-$$
>
> and so on for successive ionisation energies.

Ionisation energy

Further evidence for the existence of electrons in shells comes from a study of the ionisation energies of atoms.

Successive ionisation energies increase because the electrons are being removed from increasingly positive ions and so the attractive forces are greater. There are large jumps in ionisation energies arising from a large increase in attraction for the removed electron, which correspond to the electron being removed from a new energy level significantly closer to the nucleus. Figure 1.14 shows a graph of the logarithm of the successive ionisation energies for sodium. These energies show the electron structure of the atom; for sodium there are clearly three shells, with one electron in the outermost, a jump to the next shell with eight electrons, and then a larger one (because the positive charge from the nucleus is now much stronger) to remove the final two.

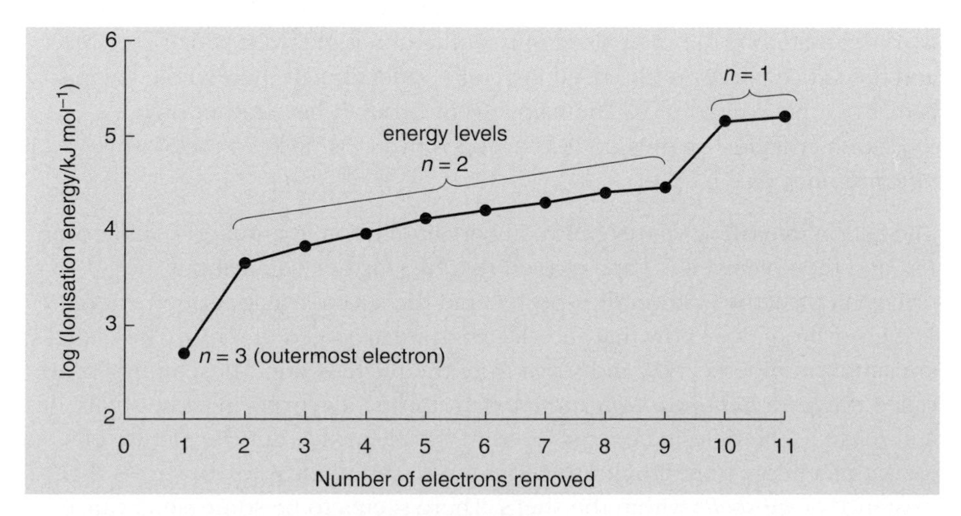

Figure 1.14 Successive ionisation energies for sodium.

QUESTION

Write the equation that represents the fifth ionisation of sodium.

Ionisation energy and the Periodic Table

Ionisation energies are important not only as evidence for energy levels, but also in determining the type of bonds which an element will form. This is dealt with fully in Chapter 3 of this book, but we must consider here the way in which ionisation energies change across the Periodic Table. The first ionisation energies for elements up to krypton, element 36, are plotted against atomic number in Figure 1.15.

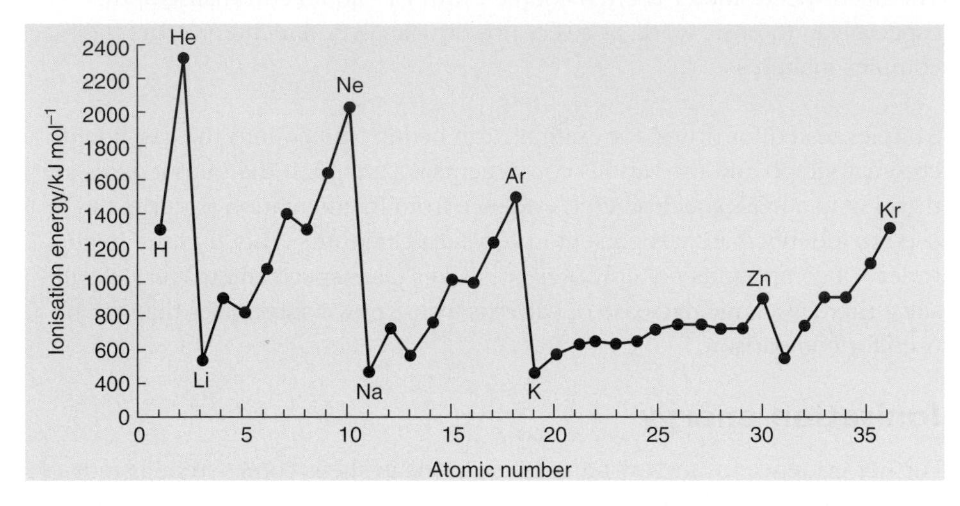

Figure 1.15 The first ionisation energies (in kJ mol^{-1}) of the elements up to krypton.

QUESTION

Sodium ions are smaller than sodium atoms; chloride ions are larger than chlorine atoms; sodium atoms are larger than chlorine atoms, but sodium ions are smaller than chloride ions. Why?

The repetition of the general pattern up to element 20 is clear. The highest points are the noble gases in Group 0, the lowest ones the alkali metals. Hydrogen is quite high at 1312 kJ mol^{-1}, and its ion is never found as such in compounds; it is invariably bonded with something else, for example water in aqueous acids to form the hydroxonium ion H_3O^+, the energy released in this process compensating for the ionisation energy.

When ions are formed, the overall energy change is often positive; this energy is given back, and more, when the ions come together to form electrostatic (ionic) bonds. The energy given out is called the **lattice energy**. The low values of ionisation energy for the alkali metals mean that the ionisation energy is compensated by the electron affinity of the non-metal with which they react and the lattice energy of the resulting compound; almost their whole chemistry is that of ionic compounds. The halogens in Group 7 have rather high ionisation energies, so they do not form positive ions. Instead they form negative ones (see p. 15).

The falls in ionisation energy between beryllium (atomic number 4) and boron (5), and then nitrogen (7) and oxygen (8), give further detail about the electronic structures. From the spectra, and the successive ionisation energies of a given atom, we know that the electrons are arranged in *shells*. These shells are simply numbered 1, 2, and so on from the nucleus out. These numbers are called the *principal quantum numbers*, *n*, for the electrons in these shells, and relate to the differing energy levels of the electrons. But the picture of ionisation energy from lithium to neon (atomic numbers 3 to 10) shows the existence of *subshells* within the shells. There seems to be some significance concerning the elements grouped as a pair, lithium and beryllium, and the two triplets: boron, carbon, nitrogen and oxygen, fluorine, neon. The full story of the discovery of the subshells is complex and not appropriate here, but in the second shell there are two subshells called the s and p; in the third shell there are three, the s, p and d. The subshells of a given shell are quite close together in energy. The s subshell can contain two electrons, the p six, and the d ten. The first shell has no subshell. So the shells and subshells up to the fourth shell (which also has an 'f' subshell) can contain the following number of electrons:

subshell	1s	2s	2p	3s	3p	3d	4s	4p	4d	4f
maximum number of electrons	2	2	6	2	6	10	2	6	10	14

There is now a further layer to the atomic onion. Each subshell has further divisions, one in the s subshell, three in the p subshell and five in the d. (You will have noticed that this is half the number of permissible electrons.) These sub-subshells are called *orbitals*, and their nature is considered further later in this chapter.

The 'Aufbau' principle

Although these arrangements arose from the mathematical theory of atomic structure, it is enough for your purposes to think of the subshells as boxes into which atoms put electrons. So we'll see how the electrons are arranged in these boxes. The 'Aufbau' or 'build-up' principle enables us to predict the electronic structure of an atom. Electrons are added to the lowest energy orbital available, one at a time, with no more than two electrons occupying any one orbital. If there are several orbitals of the same energy available (for example there are three p orbitals) then electrons enter these orbitals singly so as to be as far apart as possible. The order of filling which results is shown in Figure 1.16. It can be justified from the mathematical theory of the atom.

Each orbital can hold (more accurately can be) one or two electrons. These differ in a property called *spin*: the electrons are shown as an up or a down arrow to distinguish the different spins. The implications of electron spin are not important here.

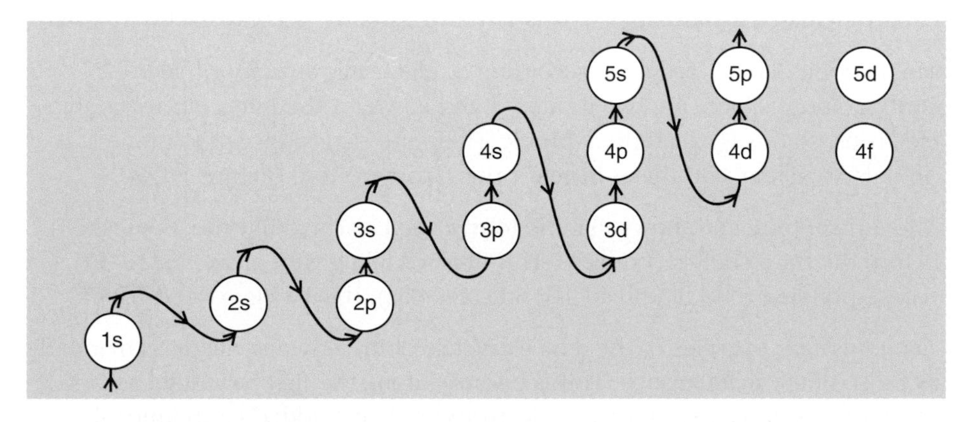

Figure 1.16 The order of filling of atomic orbitals.

Let us consider the electronic structures for the first 21 elements, and see how they relate to the structure of the Periodic Table and to the ionisation energy graph of Figure 1.15.

Using the idea of electrons in boxes, hydrogen and helium are easy; electrons go into the 1s shell.

The 1s is now full; it is very difficult to remove electrons from full shells, so the ionisation energy of helium is high. The next electron must go in to the 2s subshell, according to our picture, so lithium is

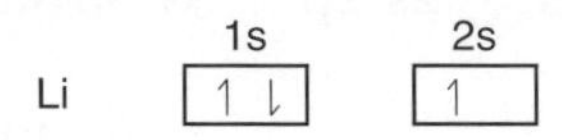

This can also be written $1s^2 2s^1$, i.e. principal quantum number, subshell, number of electrons.

The lone electron is quite easy to remove; a full shell is not being broken, and the 3+ charge on the lithium nucleus is to some extent shielded from the outer electron by the inner electrons, so the outer electron is held by an effective nuclear charge which is rather less than three (actually about +1.3). With beryllium, the next electron also goes into the 2s orbital giving: $1s^2 2s^2$. The 2s subshell is now full, so in the boron atom the 2p subshell starts to fill, giving $1s^2 2s^2 2p^1$. The 2p electron, somewhat on its own, is slightly easier to remove than is the 2s electron from boron, because subshells that are full are somewhat more stable than those which are not. Carbon has the electron structure $1s^2 2s^2 2p^2$, and nitrogen $1s^2 2s^2 2p^3$. Oxygen is $1s^2 2s^2 2p^4$:

	1s	2s	2p		
C	↑↓	↑↓	↑	↑	
N	↑↓	↑↓	↑	↑	↑
O	↑↓	↑↓	↑↓	↑	↑

It's not hard to see that the final electron will be repelled a little since it is paired up in a 2p orbital, so it is a bit easier to remove. Although the next electron introduced will also be repelled by the one already there, this effect is easily offset by the smaller size and increased nuclear charge of fluorine, so the ionisation energy increases significantly. Neon is $1s^2 2s^2 2p^6$, and the second shell is now full.

The Periodic Table is arranged according to electronic structure (electron configuration). There are two groups, 1 and 2, where the outer electrons are s electrons, so this is called the **s-block**. There are six groups where the p-orbitals are filling, and this **p-block** forms Groups 3 to 0 (Figure 1.17).

The Aufbau principle shows that the fourth shell starts to fill (the 4s subshell) before the third shell is complete. This finishes filling with elements 21–30 (corresponding rows fill the 4d and 5d subshells); this block is the **d-block**.

The ionisation energies of the d-block metals do not change nearly as markedly as those of the main groups. This is because along the first row of these elements, from scandium to zinc, electrons are being added to an inner shell (see Table 1.2, p.14) and the effect of increased nuclear charge is more or less compensated by these added electrons. So the sizes of the atoms of the d-block elements do not change enormously, and neither do their ionisation energies.

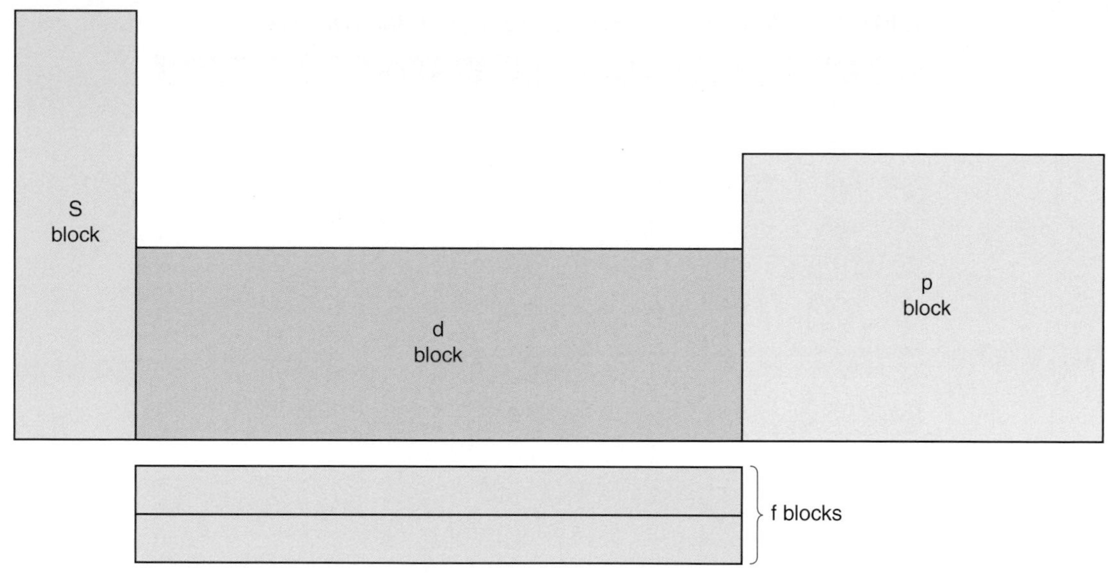

Figure 1.17 The Periodic Table is arranged in 'blocks', based on the electron configuration in the orbitals.

The electronic structures of the elements up to krypton are shown in Table 1.2. Since they can get quite long, electron configurations are often shown in abbreviated form, where the symbol of the nearest inert gas of lower atomic mass than the element of interest, is shown in square brackets; thus potassium could be written [Ar]$4s^1$.

The electronic structure is central to the chemistry of an element; the number of the electrons in the outer (valence) shell and, in some cases the shell next to the outer shell, together with the ionisation energies and electron affinities for an atom, will determine the chemistry of the element.

> **QUESTION**
>
> Write the electron configurations of the following elements, both in 1s 2s 2p . . . notation and as electrons-in-boxes: ${}_{9}F$, ${}_{14}Si$, ${}_{21}Sc$, ${}_{31}Ga$.

Atomic orbitals

We have already mentioned that the orbiting electron model of the atom was known to be in error when Bohr proposed it. The solution to this problem came in 1928, when Schrödinger and others developed *quantum mechanics*, which, amongst other things, said that the electron in an atom behaves as if it is a wave, not a particle. This is a difficult idea, and its development does not belong here, but instead of orbiting electrons we have the **atomic orbital**. This is a volume of space, often of strange shape, which is occupied by up to two electrons. This is how it is often described; in fact the orbital *is* the electron or pair of electrons. Don't try to visualise an electron turning into a wave, or existing simultaneously as a wave and a particle. In any case this wave is a mathematical construction, not a wave actually *in* anything. Instead, the orbital can be seen as either the volume in which the electron has a 95% probability of being found; or, in a much looser but quite useful way, a volume of space which has the property which might be called 'electron-ness', that is a volume within which the electron can behave as electrons do. Principally this is in formation of bonds, which form by the overlap of orbitals thus increasing the electron density between the bonded atoms. This is covered in detail in Chapter 3.

Table 1.2 *Electronic structures of the first 36 elements*

Z	Symbol	1s	2s	2p	3s	3p	3d	4s	4p
1	H	1							
2	He	2							
3	Li	2	1						
4	Be	2	2						
5	B	2	2	1					
6	C	2	2	2					
7	N	2	2	3					
8	O	2	2	4					
9	F	2	2	5					
10	Ne	2	2	6					
11	Na	2	2	6	1				
12	Mg	2	2	6	2				
13	Al	2	2	6	2	1			
14	Si	2	2	6	2	2			
15	P	2	2	6	2	3			
16	S	2	2	6	2	4			
17	Cl	2	2	6	2	5			
18	Ar	2	2	6	2	6			
19	K	2	2	6	2	6		1	
20	Ca	2	2	6	2	6		2	
21	Sc	2	2	6	2	6	1	2	
22	Ti	2	2	6	2	6	2	2	
23	V	2	2	6	2	6	3	2	
24	Cr	2	2	6	2	6	5	1	
25	Mn	2	2	6	2	6	5	2	
26	Fe	2	2	6	2	6	6	2	
27	Co	2	2	6	2	6	7	2	
28	Ni	2	2	6	2	6	8	2	
29	Cu	2	2	6	2	6	10	1	
30	Zn	2	2	6	2	6	10	2	
31	Ga	2	2	6	2	6	10	2	1
32	Ge	2	2	6	2	6	10	2	2
33	As	2	2	6	2	6	10	2	3
34	Se	2	2	6	2	6	10	2	4
35	Br	2	2	6	2	6	10	2	5
36	Kr	2	2	6	2	6	10	2	6

The shapes of the atomic orbitals for hydrogen are shown in Figure 1.18. The orbitals are assumed to be the same shape for other atoms. This is necessary because exact calculation for atoms with more than one electron is not at present possible, since the mathematical techniques do not exist. When we come to bonding we will sometimes talk of empty orbitals; this is a convenience, as you will see, but since the orbital is the electron, there is really no such thing as an empty orbital.

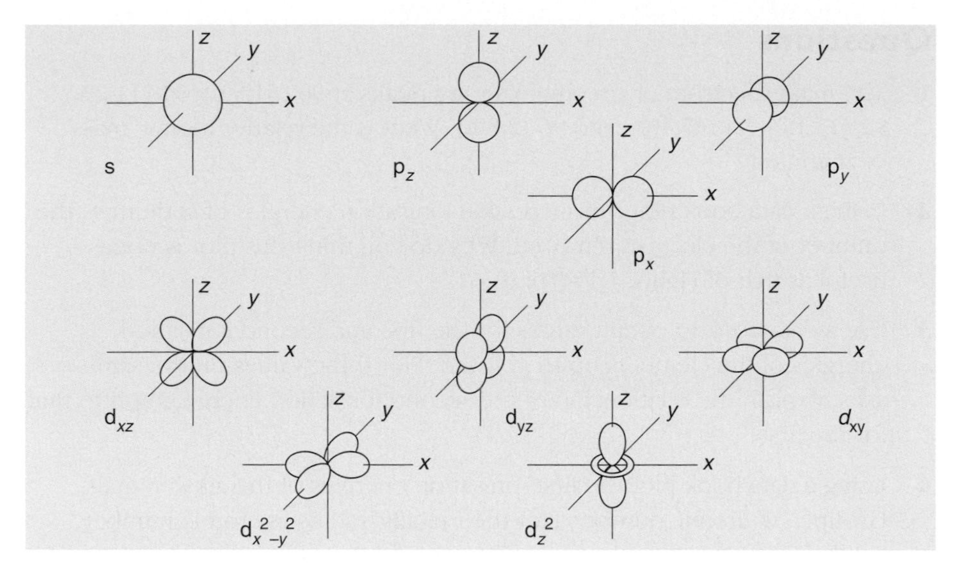

Figure 1.18 Hydrogen atomic orbitals.

Electron affinity

The electron affinity of an atom is important when deciding what sort of bonding non-metals should show, and is considered further in Chapter 3. The first electron affinity is the energy change when one mole of atoms in the gas phase each gain one electron:

$$X(g) + e^- \rightarrow X^-(g)$$

In general, this is a negative (exothermic) quantity, since the electron is attracted by the positive charge on the atom's nucleus. Some values (in kJ mol^{-1}) are given in Table 1.3.

The second electron affinity is positive, because in the reaction

$$X^-(g) + e^- \rightarrow X^{2-}(g)$$

an electron is being added to an ion which is already negative, and there is repulsion between the similar charges.

DEFINITION

The **electron affinity**, also called the **electron gain energy**, is defined as the energy change per mole for

$$X(g) + e^- \rightarrow X^-(g)$$

that is for the acquisition of electrons by gaseous atoms to form anions in the gas phase.

Table 1.3 *Some electron affinities*

Species	Electron affinity/kJ mol^{-1}
H	−72.8
C	−122.3
O	−141.1
O^-	about +798
F	−328.0
P	−72.0
S	−200.4
S^-	+640.0
Cl	−348.8
Br	−324.6
I	−295.4

Questions

1 The mass spectrum of zirconium shows peaks at 90 (51.5%), 91 (11.2%), 92 (17.1%), 94 (17.4%), and 96 (2.8%). What is the relative atomic mass of zirconium?

2 Using a data book plot the successive ionisation energies of sodium vs the number of the electron removed. Why do you think this plot is not as useful as that of Figure 1.14? (p. 9)

3 Use a data book to obtain values for the first and second ionisation energies of the elements up to krypton. Plot these values on the same axes. Explain the relationship of the second ionisation energy graph to that for the first.

4 Using a data book plot the first ionisation energies of the alkali metals, Group 1, vs atomic number. Plot the metallic radius vs atomic number for the same group, and comment on the relationship between the two graphs.

5 Using a data book to obtain the appropriate values, plot the values of (a) covalent radius, (b) ionisation energy, (c) electron affinity, vs atomic number for the halogens, Group 7. Use small graphs. Make any comments you can about the relationship between these graphs.

Further questions on background material:

6 Rutherford used a foil about 1 μm thick for the scattering experiments. The diameter of a gold atom is roughly 3×10^{-10} m. Roughly how many layers of gold atoms did the α-particles pass through?

7 The ionisation energy of hydrogen can be found from its spectrum; so can that of hydrogen-like (one-electron) species such as Li^{2+}. The spectral lines converge to a limit, which corresponds, via $E = hf$, to the ionisation energy. Plot the following frequencies of the spectral lines of Li^{2+} vs the number of the transition, and extrapolate the graph to get a value for the convergence limit. Evaluate the energy required for the process

$$Li^{2+}(g) \rightarrow Li^{3+}(g) + e^-.$$

Frequencies/10^{14} Hz: 2.20 2.63 2.77 2.84 2.88 2.90

Planck's constant $h = 6.6262 \times 10^{-34}$ J s.

2

Quantitative chemistry

The ability to calculate is central to all physical sciences; a hypothesis may look delightful when stated in *qualitative* terms, but the acid test is always whether the *quantitative* predictions that have been made stand up to examination. Calculations occur throughout chemistry, but in this chapter only reacting masses and volumes are considered, together with the use of chemical equations to represent reactions.

When solving calculation problems you should always aim to understand what is going on rather than relying on formulae learned by rote, which are easily mis-remembered or applied wrongly, and you should do as many examples as you can.

Empirical and molecular formulae

Chemists write equations for several reasons.

- They are internationally understood. A chemistry paper written anywhere in the world in any alphabet system will have recognisable equations in it.
- They are quantitative; they tell you not only which substances are involved, but also how much of each.
- They are shorter than the same information given in words. The 'word equation' is not an equation at all because it is not quantitative.

Equations contain formulae, and formulae are calculated from analytical data. When a new compound is made, it will be analysed to see how much of each element is present in it. This is now done by machine rather than by the lengthy methods necessary in the past, but you have to know how to find the empirical formula from the results.

The **empirical formula** shows the ratio of atoms present in their lowest terms, i.e. smallest numbers. Any compound having one hydrogen atom for every carbon atom will have the empirical formula CH; calculation of the **molecular formula** will need extra information, since ethyne, C_2H_2, cyclobutadiene, C_4H_4, and benzene, C_6H_6, all have CH as their empirical formula. Empirical formulae are initially found by analysing a substance for each element as a percentage by mass.

QUESTION

Find the empirical formula of the compound containing C 22.02%, H 4.59%, Br 73.39% by mass.

As an example we shall use the compound having the composition by mass 52.18% carbon, 13.04% hydrogen, and 34.78% oxygen. When finding the numbers of each atom, the different masses of the atoms must be considered. The percentage by mass is therefore divided by the relative atomic mass, i.e. 12 for carbon, 1 for hydrogen and 16 for oxygen, giving:

$$C_{\frac{52.18}{12}}\ H_{\frac{13.04}{1}}\ O_{\frac{34.78}{16}} = C_{4.35}H_{13.04}O_{2.17}$$

QUESTION

A compound contains H 1.75%, S 28.08%, O 70.17% by mass. Find its empirical formula. Given that this is also its molecular formula, write the equation for its reaction with aqueous sodium hydroxide.

QUESTION

A compound contains C 62.08%, H 10.34%, O 27.58% by mass. Find its empirical formula and its molecular formula given that its relative molecular mass is 58.

This gives the ratio for each of the atoms in moles. These numbers are now divided by the smallest to get whole-number ratios:

$$C_{\frac{4.35}{2.17}}\ H_{\frac{13.04}{2.17}}\ O_{\frac{2.17}{2.17}} = C_2H_6O$$

which is the empirical formula.

This is most usefully summarised in a table:

	C	H	O
%	52.18	13.04	34.78
Divide by A_r	52.18/12 =4.35	13.04/1 =13.04	34.78/16 =2.17
Divide by smallest	4.35/2.17 =2	13.04/2.17 =6	2.17/2.17 =1
Empirical formula		**C_2H_6O**	

Occasionally, mostly in substances containing only two atoms, this process does not give entirely whole-number answers. So the compound containing 82.76% C and 17.24% H by mass gives $CH_{2.5}$ as its empirical formula by the method above. In this case, multiply by 2 to get C_2H_5 which is the empirical formula. If such multiplication is needed, it will always be by a small whole number.

In order to get the molecular formula you need to know something else about the compound other than its percentage composition by mass. Commonly the relative molecular mass is given. For example a compound with 92.31% carbon and 7.69% hydrogen gives an empirical formula of CH, which has a relative mass of 13. If the compound has a relative molecular mass of 78, the number of CH units in each molecule is 78/13, or six. The molecule is therefore C_6H_6.

You could be given other information to convert the empirical into the molecular formula. If you find the empirical formula for the compound with 26.67% carbon, 71.11% oxygen, 2.22 % hydrogen, you will get CO_2H. If you are told that the substance is a dibasic acid, i.e. one which has two replaceable hydrogen atoms per molecule, it makes the molecular formula $C_2H_2O_4$.

Equations

You cannot avoid equations in chemistry! So use every opportunity you have to practise them, and if you cannot work them out look them up; the use of reference materials such as other textbooks or data books or journals is in any case an important part of learning how to learn. You should make the resolutions to: *never refer to any reaction in anything you write unless you give the equation for it*, and *never refer to a compound (or indeed read about one) without learning its formula*. Formulae and equations are the vocabulary and grammar of chemistry, and it is impossible to know too many.
It is pointless to give lots of examples here, since there are equations scattered throughout the book. The following principles always apply:

1 **Equations balance for mass**, so the left and right-hand sides have the same number of each type of atom. If equations need balancing, for example

$$Fe(s) + Cl_2(g) \rightarrow FeCl_3(s)$$

you cannot change the formulae. Balancing must be done by altering only the numbers in front of the formula, i.e.

$$2Fe(s) + 3Cl_2(g) \rightarrow 2FeCl_3(s)$$

2 **Equations must balance for total charge**; if there are two positives on the left, there must be two on the right. These could come from any combination of charges, as long as the total charge is the same. Thus in

$$MnO_4^- + 5Fe^{2+} + 8H^+ \rightarrow Mn^{2+} + 5Fe^{3+} + 4H_2O$$

the charges on each side are not the same, but their sum (17+) is.

Ionic equations

Ionic equations are often used in inorganic chemistry to reduce the amount of unnecessary information and to highlight the processes taking place. To turn an ordinary equation into an ionic equation, you

1 Write all soluble ionic compounds with the ions separated.
2 Write all insoluble ionic compounds and all covalent compounds in the usual manner.
3 Cross out ions which appear on both sides of the equation. These are called **spectator ions**, since they are sitting there, not involved in the reaction, while things happen around them.

An example is the reaction between aqueous solutions of sodium chloride and silver nitrate, to give a white precipitate of the insoluble silver chloride, and a solution of sodium nitrate:

$$NaCl(aq) + AgNO_3(aq) \rightarrow NaNO_3(aq) + AgCl(s)$$

Writing down the ions, we have

$$Na^+(aq) + Cl^-(aq) + Ag^+(aq) + NO_3^-(aq) \rightarrow Na^+(aq) + NO_3^-(aq) + AgCl(s)$$

Deleting the common (spectator) ions from both sides, we get the ionic equation for the reaction:

$$Ag^+(aq) + Cl^-(aq) \rightarrow AgCl(s)$$

No information that matters has been lost, since the interest lies in the precipitation of the silver chloride, not the ions that are left over.

In some cases the simplification is considerable; thus in the reaction between iron(II) sulphate and potassium manganate(VII) in the presence of sulphuric acid, the full equation is

$$2KMnO_4(aq) + 8H_2SO_4(aq) + 10FeSO_4(aq) \rightarrow$$
$$2MnSO_4(aq) + 5Fe_2(SO_4)_3(aq) + K_2SO_4(aq) + 8H_2O(l)$$

Table of common ions

Cations (positive ions)		Anions (negative ions)	
lithium	Li^+	chloride	Cl^-
sodium	Na^+	bromide	Br^-
potassium	K^+	iodide	I^-
magnesium	Mg^{2+}	nitrate	NO_3^-
calcium	Ca^{2+}	nitrite	NO_2^-
barium	Ba^{2+}	sulphate	SO_4^{2-}
aluminium	Al^{3+}	sulphite	SO_3^{2-}
chromium(III)	Cr^{3+}	carbonate	CO_3^{2-}
manganese(II)	Mn^{2+}	hydrogen carbonate	HCO_3^-
iron(II)	Fe^{2+}	chlorate(I)	OCl^-
iron(III)	Fe^{3+}	chlorate(V)	ClO_3^-
copper(II)	Cu^{2+}	cyanide	CN^-
zinc	Zn^{2+}	manganate(VII)	MnO_4^-
silver	Ag^+	chromate(VI)	CrO_4^{2-}
ammonium	NH_4^+	dichromate(VI)	$Cr_2O_7^{2-}$

Writing down the ions:

$$2K^+(aq) + 2MnO_4^-(aq) + 16H^+(aq) + 18SO_4^{2-}(aq) + 10Fe^{2+}(aq) \rightarrow 2Mn^{2+}(aq) + 10Fe^{3+}(aq) + 18SO_4^{2-}(aq) + 2K^+(aq) + 8H_2O\ (l)$$

Erasing the spectator ions leads to the ionic equation

$$MnO_4^-(aq) + 8H^+(aq) + 5Fe^{2+}(aq) \rightarrow Mn^{2+}(aq) + 5Fe^{3+}(aq) + 4H_2O(l)$$

State symbols should be included, since they help you to visualise the reaction taking place.

Reacting masses

The Avogadro constant and the mole

Equations enable you to calculate how much material to use to get a desired amount of product, whatever the scale. The essential link between the equation and the quantities of material you weigh out is the mole.

Suppose the reaction of interest is

$$Fe(s) + S(s) \rightarrow FeS(s)$$

DEFINITION

- The **Avogadro constant**, N_A, is $6.02 \times 10^{23}\ mol^{-1}$.
- One mol of any substance is 6.02×10^{23} particles of it; depending on the substance these particles may be molecules, ions or atoms.
- One mol of any substance is its relative molecular or relative atomic mass expressed in grams; this is also called the **molar mass**. (The SI symbol for the Avogadro constant is *L*.)

Iron and sulphur react on heating to give iron(II) sulphide, which is a hard, dark-grey solid. The equation shows that one atom of iron reacts with one of sulphur; and we know from tables of relative atomic masses that iron has a relative atomic mass of 56 and sulphur 32. One atom of each cannot be weighed, but since the elements react in the ratio of 56 parts of iron to 32 parts of sulphur by mass, we could use 56 g of iron and 32 g of sulphur which we can weigh out. Further, since these large masses are in the same ratio as the masses of the atoms, 56 g of iron and 32 g of sulphur contain the same number of atoms.

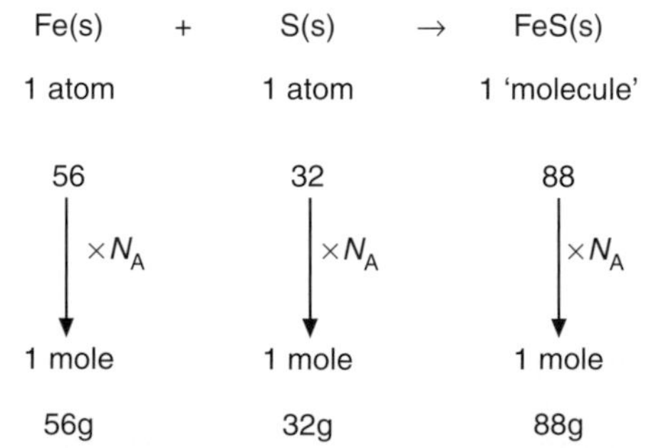

Fe(s)	+	S(s)	→	FeS(s)
1 atom		1 atom		1 'molecule'
56		32		88
↓ $\times N_A$		↓ $\times N_A$		↓ $\times N_A$
1 mole		1 mole		1 mole
56g		32g		88g

This number of atoms is called the Avogadro constant, N_A, after the Italian chemist Amadeo Avogadro, who in 1811 stated that equal volumes of all gases under the same conditions contain the same number of particles. Its value is $6.02 \times 10^{23}\ mol^{-1}$, and this number of particles defines the amount of substance called the **mole**. One mole of any substance is 6.02×10^{23} particles of it, which may be molecules or ions or atoms, depending on the substance. Such an enormous number of particles cannot be counted, but can be weighed; a mole of iron is 56 g, and that of sulphur is 32 g. A mole of any substance is that substance's atomic or molecular mass expressed in grams, and this mass is called the **molar mass**. Unlike relative atomic or molecular masses, it has units. The **number of moles of any substance** is the **amount** of it. *The word 'amount' is used in a technical sense in chemistry, with a precise meaning.*

The magnitude of N_A was first found by the German chemist Loschmidt, and in Europe it is often called Loschmidt's number. Its value is seldom needed, but the idea lies at the bottom of every reacting mass calculation. (The values of physical constants, and relative atomic masses, will always be given in any examination.)

Reacting masses

Now that you have the idea of the mole, and can see the relationship of masses of substances to the numbers of atoms or molecules involved, calculation of the amounts of any of the substances is simply an exercise in proportion, if the amount of one of them is known. The amount can then be converted into the mass.

Refer again to the iron–sulphur reaction. All calculation is done in terms of numbers of moles, that is the amount, where

$$\text{amount (number of moles)} = \frac{\text{mass of substance}}{\text{molar mass}}$$

Suppose 5.6 g of iron was used; how much sulphur is needed, and how much iron(II) sulphide is produced? The calculation is as follows:

$$Fe(s) + S(s) = FeS(s)$$

$$\text{amount of iron} = \frac{5.6\text{ g}}{56\text{ g mol}^{-1}}$$

$$= 0.1\text{ mol}$$

The number of moles of sulphur and iron(II) sulphide is the same as that of the iron, according to the equation. The mass of each substance, then, is equal to the **number of moles times the molar mass of that substance**:

	$Fe(s)$ +	$S(s)$	$\rightarrow$	$FeS(s)$
mol	0.1	0.1		0.1
mass		$0.1\text{ mol} \times 32\text{ g mol}^{-1}$ $= 3.2\text{ g}$		$0.1\text{ mol} \times 88\text{ g mol}^{-1}$ $= 8.8\text{ g}$

Suppose that 0.82 g of iron had been used at first.

	$Fe(s)$ +	$S(s)$	$\rightarrow$	$FeS(s)$
mol	$\frac{0.82\text{ g}}{56\text{ g mol}^{-1}}$ $= 0.0146\text{ mol}$			
mass		$0.0146\text{ mol} \times 32\text{ g mol}^{-1}$ $= 0.467\text{ g}$		$0.0146\text{ mol} \times 88\text{ g mol}^{-1}$ $= 1.285\text{ g}$

QUESTION

What mass of sulphur is required to make 1 tonne (1000 kg) of sulphuric acid, H_2SO_4?

QUESTION

Lead(IV) oxide reacts with concentrated hydrochloric acid according to:

$$PbO_2(s) + 4HCl(aq) \rightarrow PbCl_2(s) + Cl_2(g) + 2H_2O$$

(a) What mass of lead chloride would be obtained from 37.2 g of PbO_2?
(b) What mass of HCl would be required?

Consider now a calculation where the proportions that are reacting are not 1:1. The reaction of sodium carbonate with hydrochloric acid is

$$Na_2CO_3(s) + 2HCl(aq) \rightarrow 2NaCl(aq) + H_2O(l) + CO_2(g)$$

To find the masses involved the molar masses of the compounds must be found. It is a good idea to write out the addition in full; if this is not done, it is impossible to tell whether errors are chemical or arithmetical. For each substance in the reaction, then, the molar masses are:

Na_2CO_3: $(2 \times 23) + 12 + (3 \times 16) = 46 + 12 + 48 = 106\,g\,mol^{-1}$

HCl: $1 + 35.5 = 36.5\,g\,mol^{-1}$

NaCl: $23 + 35.5 = 58.5\,g\,mol^{-1}$

H_2O: $(2 \times 1) + 16 = 18\,g\,mol^{-1}$

CO_2: $12 + (2 \times 16) = 44\,g\,mol^{-1}$

Suppose we are to find the various masses that react with or are produced from 7.3 g of sodium carbonate.

	$Na_2CO_3(s)$	+ $2HCl(aq)$	$\rightarrow 2NaCl(aq)$	+ $H_2O(l)$	+ $CO_2(g)$
mol	$\frac{7.3\ g}{106\ g\ mol^{-1}} = 0.0690$	0.138	0.138	0.0690	0.0690

mass

HCl $0.138\,mol \times 36.5\,g\,mol^{-1} = 5.04\,g$;

NaCl $0.138\,mol \times 58.5\,g\,mol^{-1} = 8.07\,g$;

H_2O $0.069\,mol \times 18\,g\,mol^{-1} = 1.24\,g$;

CO_2 $0.069\,mol \times 44\,g\,mol^{-1} = 3.04\,g$

Use of concentrations: volumetric analysis

Much quantitative analysis (i.e. analysing the amount of substance present) is performed using reactions between two substances in solution, the volumes of both solutions and the concentration of one of them being accurately known when the reaction is complete. The concentration of the other solution can then be found. This technique is **volumetric analysis**, or **titration**.

Remember, do not rely on mathematical formulae. Instead work in moles, and be aware of what you are doing.
Solution concentrations are usually given in mol dm^{-3} or g dm^{-3}, in both cases the "dm^{-3}" referring to the *solution*, and not to the added solvent. To make a molar solution of a substance, one mole of it is weighed out, and water added until the volume of the solution is 1 dm^3.

Acid/base titrations

In these titrations acid and base are reacted in the presence of a suitable indicator.

(The suitability of indicators is covered in *Periodicity, Quantitative Equilibrium and Functional Group Chemistry*). This may be done to find the purity of a substance, say, or to produce a **standard solution** for use in another titration.

A standard solution is one which can be made of known concentration by weighing out the solute concerned. The solute is called a **primary standard**, and must:

1. be available commercially in a high state of purity
2. be stable over long periods of time
3. not decompose when dissolved in water
4. not be volatile, so losses due to evaporation during weighing do not occur
5. not absorb water or carbon dioxide from the atmosphere.

A good primary standard alkali is anhydrous sodium carbonate; an acidic one is the solid strong acid, sulphamic acid, H_2NSO_3H.

(Volumetric calculations will be found only in Paper 3B of the Edexcel specification; in every case equations for the reactions will be given.)

Example 1

A solution of sodium carbonate contains 12.5 g of the anhydrous salt in 1 dm^3 of solution. When 25.0 cm^3 of this solution was titrated with a solution of hydrochloric acid using methyl orange indicator, 23.45 cm^3 of the acid was required. What was the concentration of the acid?

You should first write the equation for the reaction:

$$Na_2CO_3(s) + 2HCl(aq) \rightarrow 2NaCl(aq) + H_2O(l) + CO_2(g)$$

Now find the concentration of the sodium carbonate solution. Anhydrous sodium carbonate has a molar mass of 106 g mol^{-1}. The concentration of the solution is therefore

$$\frac{12.5\,g\,dm^{-3}}{106\,g\,mol^{-1}} = 0.118\,mol\,dm^{-3}$$

If you know the volume of solution used and its concentration, then

amount of solute = volume × concentration

The amount of sodium carbonate used is, therefore,

$$0.025\,dm^3 \times 0.118\,mol\,dm^{-3} = 0.00295\,mol$$

From the equation, 1 mol of Na_2CO_3 requires 2 mol HCl, so

amount of HCl used = 0.00295 mol × 2 = 0.00590 mol

This was contained in 23.5 cm^3 of solution, so the concentration of HCl is therefore

$$\frac{\text{moles of solute}}{\text{volume}} = \frac{0.00590\,mol}{0.0235\,dm^3} = 0.251\,mol\,dm^{-3}$$

QUESTION

A sample of pure anhydrous sodium carbonate weighing 1.00 g was dissolved in water and the volume made to 200 cm^3. Portions of 25.0 cm^3 of this solution were titrated with hydrochloric acid solution of concentration 0.120 mol dm^{-3}. What volume was required?

One of the greatest difficulties faced in understanding calculations is when their author fails to also write in words what is being done; a load of numbers is dumped on the page, with no connections or explanation. It is worth remembering that a lot of credit can often be given for calculations which have the wrong answer, if the reader can see, from what has been written, that many of the principles used in the work are correct.

Example 2

Solutions have frequently been diluted to obtain the solution which is titrated, but the question concerns the original solution.

A solution of sulphuric acid was made by pipetting $1.00\,cm^3$ of concentrated sulphuric acid into a $500\,cm^3$ graduated flask, and making up to the mark with pure water. The solution was well mixed, and $25.0\,cm^3$ portions titrated with sodium hydroxide solution of concentration $0.100\,mol\,dm^{-3}$, using phenolphthalein indicator. An average volume of $19.8\,cm^3$ of sodium hydroxide was needed for complete neutralisation. Find the concentration of the original acid.

$$H_2SO_4(aq) + 2NaOH(aq) \rightarrow Na_2SO_4(aq) + 2H_2O(l)$$

$$\text{amount of NaOH} = 0.0198\,dm^3 \times 0.1\,mol\,dm^{-3} = 1.98 \times 10^{-3}\,mol$$

Since 1 mol sulphuric acid requires 2 mol NaOH,

$$\text{amount of } H_2SO_4 = \frac{1.98 \times 10^{-3}\,mol}{2} = 9.9 \times 10^{-4}\,mol$$

This is the amount of acid in $25.0\,cm^3$ of diluted solution; the concentration of acid is therefore

$$9.9 \times 10^{-4}\,mol \times \frac{500}{25} \times \frac{1000}{1} = 19.8\,mol\,dm^{-3}$$

The factor of 500/25 arises because $25\,cm^3$ was taken from $500\,cm^3$, and the 1000/1 factor because that diluted solution was made from $1\,cm^3$ of the concentrated acid, whereas the answer is required for $1000\,cm^3$ of the concentrated acid.

Example 3

Some titration problems concern substances which are insoluble, for which a slightly different strategy is required. This example demonstrates the technique of **back-titration**, used here to find the purity of a sample of chalk. Since chalk is insoluble in water, a solution cannot be used. So it is reacted with a known excess of acid, the acid remaining then being titrated with standard alkali.

A sample of 1.50 g of chalk was reacted with $50.0\,cm^3$ of hydrochloric acid of concentration $1.00\,mol\,dm^{-3}$, which is an excess. When reaction had ceased, the solution was transferred quantitatively to a $250\,cm^3$ graduated flask, and made to the mark with pure water. Portions of 25.0 cm^3 of the well mixed solution were titrated with sodium hydroxide solution of concentration 0.100 mol dm^{-3}, using screened methyl orange indicator; on average a volume of 24.5 cm^3 was required. What is the percentage purity of the chalk?

$$CaCO_3(s) + 2HCl(aq) \rightarrow CaCl_2(aq) + H_2O(l) + CO_2(g)$$

$$HCl(aq) + NaOH(aq) \rightarrow NaCl(aq) + H_2O(l)$$

$$\text{Molar mass of } CaCO_3 = 40 + 12 + (3 \times 16) = 100\,g\,mol^{-1}$$

The titration determines the amount of hydrochloric acid remaining after the chalk has reacted.

$$\begin{aligned}\text{amount of NaOH} &= \text{amount of HCl unreacted}\\ &= 0.0245\ dm^3 \times 0.100\,mol\ dm^{-3}\\ &= 2.45 \times 10^{-3}\ mol\ in\ 25.0\ cm^3\end{aligned}$$

$$\begin{aligned}\text{Thus total amount of HCl unreacted} &= 2.45 \times 10^{-3}\ mol \times \frac{250}{25}\\ &= 0.0245\,mol\end{aligned}$$

$$\begin{aligned}\text{Original amount of HCl taken} &= 0.050\ dm^3 \times 1.00\ mol\ dm^{-3}\\ &= 0.050\,mol\end{aligned}$$

Thus amount of HCl used to react with the $CaCO_3$

$$\begin{aligned}&= (0.050 - 0.0245)\ mol\\ &= 0.0255\,mol\end{aligned}$$

$$\text{Therefore the amount of } CaCO_3 = \frac{0.0255\ mol}{2} = 0.0128\,mol$$

$$\text{Thus the mass of } CaCO_3 = 0.0128\,mol \times 100\,g\,mol^{-1} = 1.280\,g$$

$$\begin{aligned}\%\ \text{purity of the } CaCO_3 &= \frac{\text{mass of } CaCO_3 \text{ in sample}}{\text{mass of sample}} \times 100\\ &= \frac{1.280\,g \times 100}{1.500\,g} = 85.3\%\end{aligned}$$

Example 4

Sometimes a substance undergoes a reaction to form another substance which is then titrated. Methanal, HCHO, reacts with ammonium salts to give acid, which can then be titrated directly with base. This is a much simpler method than back-titration (Example 3) but should not be used with ammonium chloride since the hydrochloric acid produced can react with methanal to give an extremely toxic compound.

Ammonium nitrate weighing 3.205 g was dissolved in water in a graduated flask, and the volume made to 250 cm^3 with pure water. Portions of 25.0 cm^3 of the solution were treated with 5 cm^3 of 40% aqueous methanal solution, and

QUESTION

A solution of hydrochloric acid of volume 25.0 cm^3 was pipetted onto a piece of marble, which is calcium carbonate. When all action had ceased, 1.30g of the marble had dissolved. Find the concentration of the acid, assuming the marble is pure $CaCO_3$.

allowed to stand for a few minutes. The mixture was then titrated with 0.1 mol dm^{-3} sodium hydroxide solution using phenolphthalein indicator; on average a volume of 37.5 cm^3 was required . Find the percentage purity of the ammonium nitrate.

The reaction between methanal and the ammonium salt is

$$4NH_4NO_3(aq) + 6HCHO(aq) \rightarrow 4HNO_3(aq) + (CH_2)_6N_4(aq) + 6H_2O(l)$$

The organic substance $(CH_2)_6N_4$ is hexamethylene tetramine, and is neutral. Thus one mole of ammonium salt gives one mole of acid. The titration reaction is

$$HNO_3(aq) + NaOH(aq) \rightarrow NaNO_3(aq) + H_2O(l)$$

Molar mass of NH_4NO_3 = 14 + 4 + 14 + (3 × 16) = 80 g mol^{-1}

amount NaOH = 0.0375 dm^3 × 0.100 mol dm^3 = 3.75 × 10^{-3} mol

= mol HNO_3 = amount NH_4NO_3 in 25 cm^3 of solution.

Thus the total amount of ammonium nitrate is

$$3.75 \times 10^{-3}\ \text{mol} \times \frac{250}{25} = 0.0375\ \text{mol}$$

and the mass is 0.0375 mol × 80 g mol^{-1} = 3.00 g

$$\text{Therefore, \% purity of the salt} = \frac{3.00\ \text{g} \times 100}{3.205\ \text{g}} = 93.6\%.$$

Example 5

Ammonium salts can also be determined by back-titration. The salt is heated with excess sodium hydroxide solution, which liberates ammonia (it is the qualitative test for ammonium ions):

$$NH_4^+(aq) + OH^-(aq) \rightarrow NH_3(g) + H_2O(l)$$

The ammonia is absorbed in a known amount (which must be an excess) of acid, and the unreacted acid is titrated with sodium hydroxide solution.

Ammonium sulphate weighing 1.70 g was heated with excess sodium hydroxide solution, and the ammonia liberated was absorbed in 50.0 cm^3 of hydrochloric acid of concentration 1.00 mol dm^{-3}. This solution was transferred quantitatively to a graduated flask, and made to exactly 250 cm^3 with pure water. Portions of 25.0 cm^3 of this solution were titrated with sodium hydroxide solution of concentration 0.100 mol dm^{-3}, using methyl orange indicator. An average of 26.4 cm^3 was required. Find the percentage purity of the ammonium sulphate.

$$NH_3(aq) + HCl(aq) \rightarrow NH_4Cl(aq)$$

$$HCl(aq) + NaOH(aq) \rightarrow NaCl(aq) + H_2O(l)$$

amount of NaOH = 0.0264 dm^3 × 0.1 mol dm^{-3} = 2.64 × 10^{-3} mol
= amount of HCl remaining in 25 cm^3

Thus total amount of HCl remaining $= 2.64 \times 10^{-3}\ \text{mol} \times \frac{250}{25}$
$= 0.0264\ \text{mol}$

Original amount of HCl $= 0.05\ \text{dm}^3 \times 1.00\ \text{mol dm}^{-3} = 0.05\ \text{mol}$

Amount of HCl used = amount of NH_3 absorbed
$= (0.050 - 0.0264)\ \text{mol}$
$= 0.0236\ \text{mol}$

1 mol ammonium sulphate gives 2 mol NH_3:

$(NH_4)_2SO_4 + 2OH^- \rightarrow 2NH_3 + 2H_2O + SO_4^{2-}$

Therefore, amount of ammonium sulphate (molar mass $= 132\ \text{g mol}^{-1}$)
originally present $= \frac{0.0236\ \text{mol}}{2} = 0.0118\ \text{mol}$

Therefore, mass of ammonium sulphate $= 0.0118\ \text{mol} \times 132\ \text{g mol}^{-1}$
$= 1.56\ \text{g}$

% purity of ammonium sulphate $= \frac{1.56\ \text{g}}{1.70\ \text{g}} \times 100 = 91.8\%$

Example 6

The variety of volumetric calculations is very large. The problem in Example 5 could, for instance, have asked for the percentage nitrogen in the ammonium sulphate sample. Since the amount of ammonia is known, the mass of nitrogen atoms present can easily be found.

amount of NH_3 $= 0.0236\ \text{mol}$ = amount of N

mass of N $= 0.0236\ \text{mol} \times 14\ \text{g mol}^{-1} = 0.330\ \text{g}$

% N in the ammonium sulphate $= \frac{0.330\ \text{g} \times 100}{1.700\ \text{g}} = 19.4\%$

Example 7

Some salts crystallise from aqueous solution with water incorporated in the crystal structure; water of crystallisation. An example is $CuSO_4.5H_2O$. The amount of water of crystallisation can often be found by titration.

A solution of hydrated sodium carbonate (washing soda) $Na_2CO_3.xH_2O$ contained 4.00 g of the crystals on 250 cm^3 of solution. 25.0 cm^3 of this solution needed 22.38 cm^3 of 0.125 mol dm^{-3} HCl for complete neutralisation. Calculate x.

The way of doing this is to find the amount of sodium carbonate from the titration values; then to find the molar mass of the hydrated salt; then to subtract from this the molar mass of anhydrous sodium carbonate, hence finding x.

$Na_2CO_3 + 2HCl \rightarrow 2NaCl + CO_2 + H_2O$

amount of HCl used $= 0.022\,38\ dm^3 \times 0.125\ mol\ dm^{-3}$
$= 2.798 \times 10^{-3}\ mol$

thus amount of sodium carbonate in $25.0\ cm^3$ solution $= 1.399 \times 10^{-3}$ mol
thus total amount of sodium carbonate used $= 1.399 \times 10^{-2}$ mol
molar mass of the hydrate is therefore $4.00\ g/1.399 \times 10^{-2}$ mol
$= 286\ g\ mol^{-2}$
molar mass of anhydrous sodium carbonate $= 106\ g\ mol^{-1}$
thus 1 mole of hydrate contains (286 – 106) g of water
therefore $x = 180\ g\ mol^{-1}/18\ g\ mol^{-1} = 10$

Sodium carbonate decahydrate $Na_2CO_3.10H_2O$ is washing soda. It loses water to the air on storage, so the value of x can often be less than 10.

Other titrations

Although explained in detail in *Transition Metals, Quantitative Kinetics and Applied Organic Chemistry*, redox titrations do not differ in principle from acid/base titrations. They do not appear as part of unit test 1.

Example 8

Potassium manganate(VII) reacts with iron(II) ions according to the equation:

$$MnO_4^-(aq) + 5Fe^{2+}(aq) + 8H^+(aq) \rightarrow Mn^{2+}(aq) + 5Fe^{3+}(aq) + 4H_2O(l)$$

Solutions of iron(II) are titrated with potassium manganate(VII) which, being intensely purple, is its own indicator.

Impure iron weighing 1.650 g was dissolved in $100\ cm^3$ of dilute sulphuric acid in a graduated flask, and the solution was made to $250\ cm^3$ with pure water. Portions of $25.0\ cm^3$ of this iron(II) solution were titrated with potassium manganate(VII) solution of concentration $0.0200\ mol\ dm^{-3}$, $28.5\ cm^3$ being required on average. Find the percentage purity of the iron.

$$Fe(s) + H_2SO_4(aq) \rightarrow FeSO_4(aq) + H_2(g)$$

Amount of $MnO_4^- = 0.0285\ dm^3 \times 0.0200\ mol\ dm^{-3} = 5.70 \times 10^{-4}$ mol

Since 1 mol of MnO_4^- reacts with 5 mol of Fe^{2+} according to the equation,

amount of Fe^{2+} in $25\ cm^3$ of solution $= 5 \times 5.70 \times 10^{-4}$ mol
$= 2.85 \times 10^{-3}$ mol

Thus the total amount of $Fe^{2+} = 2.85 \times 10^{-3}\ mol \times \dfrac{250}{25} = 0.0285$ mol

Therefore, mass of Fe $= 0.0285\ mol \times 56\ g\ mol^{-1} = 1.596\ g$

and % Fe in the sample $\dfrac{1.596\ g}{1.650\ g} \times 100 = 96.7\%$

QUESTION

A solution of $50.0\ cm^3$ of hydrogen peroxide was diluted to $1\ dm^3$. Portions of $25.0\ cm^3$ of this solution were acidified with dilute sulphuric acid, and titrated with $0.0200\ mol\ dm^{-3}$ $KMnO_4$ solution; $23.9\ cm^3$ was required. Find the concentration of the original hydrogen peroxide.

$$2MnO_4^- + 5H_2O_2 + 6H^+ \rightarrow 2Mn^{2+} + 5O_2 + 8H_2O$$

Problems involving gas volumes

As has already been mentioned the Italian chemist Avogadro proposed that equal volumes of all gases at the same temperature and pressure contain the same number of particles. The molar volume of a gas is the most useful quantity, i.e. it is the volume, at a specified temperature and pressure, which contains 6.02×10^{23} particles of the gas. For noble (inert) gases these are atoms; for other substances they are molecules. The precise definition uses 0 °C and 1 atm pressure, where the volume is 22.414 dm^3, but no principle is lost if the value is taken as **24 dm^3 at room temperature and pressure**. The appalling vagueness of this statement within a quantitative science will not be lost on you, so the statement that the molar volume of any gas at the temperature and pressure of the experiment is 24 dm^3, is preferable. You don't need to know what the temperature and pressure actually are. This statement is included implicitly in the examples which follow. The result of the Avogadro statement is that gases react in volumes which are proportional to the mole ratios in the equation.

Example 9

Propane, C_3H_8, burns in oxygen to give carbon dioxide and water. Calculate the volumes of oxygen needed to burn 10 cm^3 of propane and of CO_2 produced.

The volumes are in the proportions stated in the equation:

$C_3H_8(g)$ +	$5O_2(g)$ =	$3CO_2(g)$ +	$4H_2O(l)$
1 vol	5 vol	3 vol	
10 cm^3	50 cm^3	30 cm^3	

Example 10

A weighed piece of marble is reacted with hydrochloric acid, and is dried and reweighed when all action has ceased. The mass loss was 2.33 g; what volume of carbon dioxide was evolved?

$$CaCO_3(s) + 2HCl(aq) \rightarrow CaCl_2(aq) + CO_2(g) + H_2O(l)$$

$$\text{amount of } CaCO_3 = \frac{2.33\,\text{g}}{100\,\text{g mol}^{-1}} = 0.0233\,\text{mol}$$

Thus amount of CO_2 produced = 0.0233 mol

Therefore **volume of CO_2 = 0.0233 mol × 24 $dm^3\,mol^{-1}$ = 0.56 dm^3.**

Questions

1 A compound containing 85.71% C and 14.29% H by mass has a relative molecular mass of 56. Find its molecular formula.

2 When copper(II) nitrate is heated, it decomposes according to:

$$2Cu(NO_3)_2(s) \rightarrow 2CuO(s) + 4NO_2(g) + O_2(g)$$

When 20.0 g of copper(II) nitrate is decomposed what mass of copper(II) oxide would be produced? If this CuO is converted to copper, what mass would be produced?

QUESTION

What mass of phosphorus is required to make 200 cm^3 of phosphine, PH_3, via the reaction:

$P_4(s) + 3NaOH(aq) + 3H_2O(l) \rightarrow 3NaH_2PO_4(aq) + PH_3(g)$

QUESTION

10.0 cm^3 of a hydrocarbon C_4H_x reacts with an excess of oxygen at 150°C and 1 atm pressure. The products occupy a volume 10 cm^3 greater than the reactants at this temperature and pressure. Find x.

3 A blast furnace could produce about 700 tonnes of iron a day.

(a) How much iron(III) oxide would be consumed?

(b) Assuming coke is pure carbon, how much coke would be needed to produce the necessary carbon monoxide?

$$Fe_2O_3(s) + 3CO(g) \rightarrow 2Fe(l) + 3CO_2(g)$$
$$2C(s) + O_2(g) \rightarrow 2CO(g)$$

4 A sample of impure ammonium sulphate weighing 1.852 g was dissolved in water and made to 250 cm^3. Portions of 25.0 cm^3 of this solution were treated with excess sodium hydroxide and distilled. This liberated ammonia which was absorbed in 50.0 cm^3 of 0.106 mol dm^{-3} hydrochloric acid (an excess). When distillation was complete, the acid remaining required 25.5 cm^3 of 0.102 mol dm^{-3} sodium hydroxide for neutralisation. Find the percentage purity of the ammonium nitrate.

5 Dry hydrogen chloride gas of volume 120 cm^3 was absorbed in water and the resulting solution made to a volume of 100 cm^3. Find the concentration of this solution, and the volume of 0.125 mol dm^{-3} sodium hydroxide solution required to exactly neutralise it. (Molar volume of gas at the temperature and pressure of the experiment is 24 dm^3.)

3 Bonding and structure

Bonds

Atoms would not be of much interest if they weren't able to join together; there would be no chemistry, after all. Their joining is not random, though, and any model of bonding must be able to explain why many things cannot easily be pulled apart, yet people do not permanently bond to the chair on which they are sitting.

A fundamental feature of chemistry is that the bulk properties of substances, whether they are hard or soft, soluble or not, conductors or non-conductors of electricity, are explicable in terms of the bonding between atoms and molecules. The properties depend on the nature of the bonds, and on exactly how the bonds are distributed throughout the material. Bonding occurs where atoms or molecules can rearrange their electrons to give lower energy arrangements of electrons and nuclei than were present before bonding took place.

Until now you have probably divided bonds into two types, the ionic bond where positively and negatively charged ions are held together by the attraction of their opposite charges, or the covalent bond where electrons are shared between the bonded atoms. Most compounds have some of the characteristics of both, though tending in major properties towards one or the other. Thus sodium chloride is almost completely ionic (Figure 3.1), but lithium iodide shows some covalent characteristics; chlorine or hydrogen molecules are completely covalent, but hydrogen fluoride has electrons which are very unequally shared and the bond shows some polarity or separation of charge. Some compounds have both sorts of bond; NH_4NO_3 has ammonium and nitrate ions held ionically, but the atoms in the ions themselves are covalently bonded.

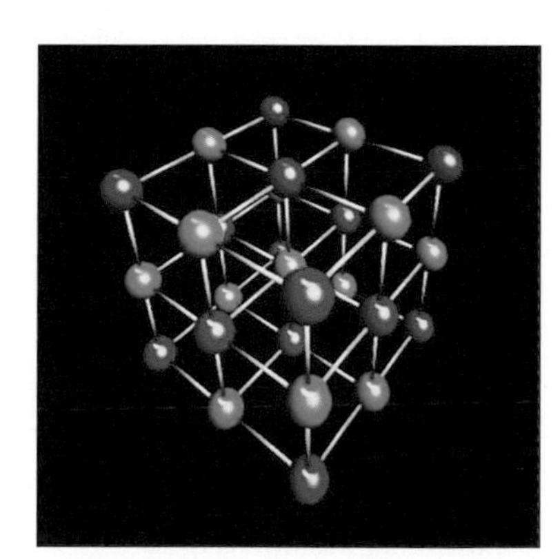

Figure 3.1 The NaCl lattice: its properties are determined by its bonding.

The present approach to bonding will ask firstly how the predominant type may be predicted, and secondly how far the compound is likely to deviate from that type. The notion that metal bonds to non-metal with ionic bonds and non-metal with non-metal by covalent bonds was useful at GCSE, but must now be improved to a more informed approach.

The ionic bond

Ions

Ionic bonds are formed between particles which have a net electrical charge. Positive ions are called **cations**, since they are attracted to the negative cathode during electrolysis; negative ions are called **anions**, and migrate to the anode on electrolysis. Either sort of ion can contain one or more atoms.

Definition

An ionic bond is formed by the attraction between oppositely charged ions in a crystal lattice.

Simple cations are metal atoms which have lost one or more electrons, e.g. K^+, Ca^{2+}. They have more protons than electrons, hence the positive charge.

The hydration of metal ions

In solution metal ions attract water molecules. Originally this was thought to be the result only of the polar water molecules being attracted to the charge on the metal ion, with the number of water molecules involved being somewhat variable. Now there is much evidence to suggest that complex ions of the form $[M(H_2O)_6]^{x+}$ are formed by many metal cations, not only those of the transition metals for which such complexes have long been known.

The metal ion is joined to the six water **ligands** by **dative covalent bonds**, that is bonds where both electrons come from one atom. The water molecules donate a lone pair of electrons to empty orbitals on the metal ion and form octahedral complexes. Examples include $[Mg(H_2O)_6]^{2+}$ (colourless), $[Cu(H_2O)_6]^{2+}$ (pale blue) and $[Mn(H_2O)_6]^{2+}$ (very pale pink). The ions formed by transition metals are usually coloured.

Polyatomic cations have several atoms bonded covalently, the whole structure having a positive charge, e.g. NH_4^+.

Anions have more electrons than protons. Simple anions are those of non-metals, e.g. Cl^-, O^{2-}. Complex anions exist where the groups around the central metal ion are themselves negative, e.g. the iron(II) complex $[Fe(CN)_6]^{4-}$. Polyatomic anions are common, derived from acids by the loss of one or more hydrogen ions, and include SO_4^{2-}, NO_3^-, CH_3COO^-, MnO_4^- and $Cr_2O_7^{2-}$.

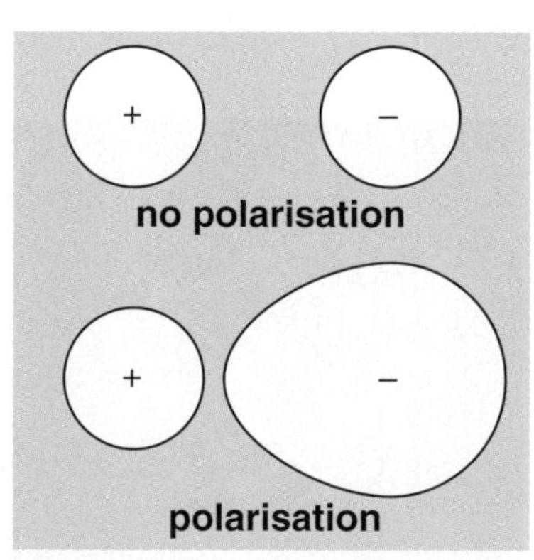

Figure 3.2 The ionic model has no polarisation; deviations arise because polarisation gives some covalent character.

Features favouring ionic bonding

Ionic bonds (Figure 3.2) are favoured if

- the metal has a low ionisation energy
- the non-metal has a high electron affinity (electron gain energy)
- the metal forms large ions of low charge
- the non-metal forms small ions of low charge.

The extent to which the bonds formed are ionic depends on the effect which the cation, generally the smaller ion, has on the anion. Small cations of high charge have a high charge density, and this may be able to distort the electron cloud around the anion so that there is some degree of electron sharing between the cation and anion. This is **polarisation**; small highly charged cations are very *polarising*, large anions are *polarisable* (Figure 3.2).

BACKGROUND

Lattice energy

Lattice energy, ΔH_{lat} **is the energy change per mole for the process**

$$M^+(g) + X^-(g) \rightarrow MX(s)$$

The ionic model for crystals enables the **lattice energy** to be calculated. This is a measure of the strength of the bonding in an ionic substance and it is relevant when considering, for example, the solubility of an ionic compound. The calculated value assumes no polarisation, so a comparison between this and the experimental value will give an estimate of the extent of covalence within the crystal. The presence of covalence increases the lattice energy.

Consider the effect of changing the size of the cation while keeping the anion the same. As the cation gets larger, its charge density decreases and it becomes less polarising. Therefore the extent of covalence will decrease and the difference between calculated and experimental ΔH_{lat} will decrease. Table 3.1 shows this effect for the chlorides of Group 2. The larger charge density of Mg^{2+} increases its polarising power, and hence its covalence, in its compounds, compared with the rest of the group.

Table 3.1 *The lattice energies of Group 2 chlorides*

Chloride	Cation radius/pm	ΔH_{lat}/kJ mol^{-1}		
		Experimental	Calculated	Difference
$MgCl_2$	72	–2526	–2326	200
$CaCl_2$	100	–2258	–2223	35
$SrCl_2$	113	–2156	–2127	29
$BaCl_2$	136	–2056	–2033	23

QUESTION

State with reasons, which of each of the following pairs of compounds would show a greater degree of covalent character:
(a) LiCl and CsCl; (b) $MgCl_2$ and $BaCl_2$; (c) NaCl and $AlCl_3$; (d) ClF_3 and IF_3.

Consider now the effect of changing the size of the anion. As the anions increase in size, they become more polarisable since the outer electrons are further from the nucleus, less tightly held, and more prone to distortion. Similar data to that shown in Table 3.1 is given in Table 3.2, for the halides of magnesium. The more polarisable iodide ion leads to greater covalence in MgI_2 compared with MgF_2, as shown in Table 3.2.

Table 3.2 *The lattice energies of magnesium halides*

Magnesium halide	Anion radius/pm	ΔH_{lat}/kJ mol^{-1}		
		Experimental	Calculated	Difference
MgF_2	133	–2957	–2913	44
$MgCl_2$	180	–2526	–2326	200
$MgBr_2$	195	–2440	–2097	343
MgI_2	215	–2329	–1944	385

QUESTION

Criticise, by considering the energy changes involved, the statement that 'when sodium and chlorine react they do so because the ions that result have octets of electrons.'

Lastly, consider the effect of cationic charge. The greater the charge, the more polarising the cation. The difference between the experimental and calculated values of ΔH_{lat} for NaCl is 10 kJ mol^{-1}; for magnesium chloride it is 200 kJ mol^{-1}. Aluminium chloride is a covalent solid, and silicon tetrachloride is a covalent liquid at room temperature, since in these cases the differences are even larger.

The covalent bond

The sharing of electrons

Covalent bonds are formed by sharing electron pairs, one pair to a bond. For most covalent substances, this is so that hydrogen obtains two electrons in its outer shell and other atoms obtain eight (the so-called '**octet rule**'), though atoms beyond the second period (Li to Ne) can 'expand the octet' and form more than the four bonds which eight electrons allow. SF_6, where sulphur has 12 outer shell electrons, is one such compound. Only those elements which have d orbitals available for bonding can form more than four bonds.

The **dative covalent bond** is formed where both electrons in the bond come from the same atom. Examples include the ammonium ion, NH_4^+, and the hydroxonium ion H_3O^+, where the bonds, once formed, are no different from the other bonds in the ion; and in metal complexes, such as $[Mg(H_2O)_6]^{2+}$ between the ligand and the central metal ion (Figure 3.3).

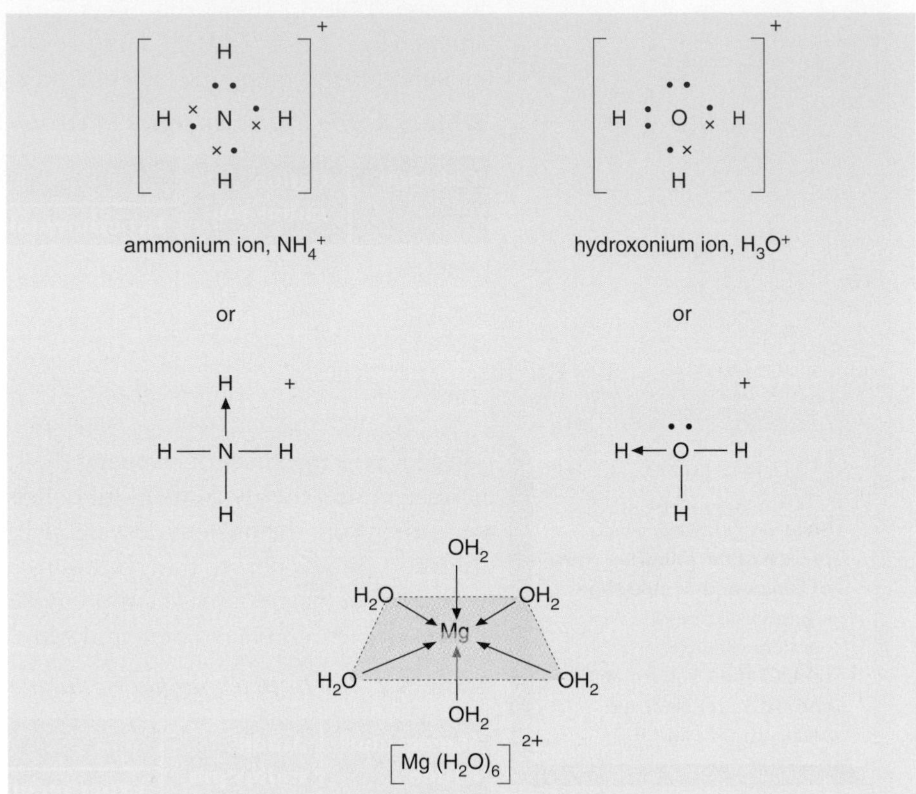

Figure 3.3 Examples of dative covalent bonds. Both electrons in the bond come from the same atom.

Orbitals and covalent bonds

A covalent bond is formed by overlap of atomic orbitals so that the electron density increases between the bonded atoms. A non-dative bond is formed by two one-electron orbitals overlapping to give a two-electron bonding orbital; a dative bond arises from a two-electron orbital donating electron density into an 'empty' orbital on the accepting atom. Have a look at the orbital shapes again (Figure 1.18) before continuing.

When the orbitals overlap they give **molecular orbitals** where the electron density extends over at least two atoms. Benzene has orbitals extending over six atoms, and graphite has orbitals over vast numbers in each sheet of atoms (see Bonding and properties, later in this chapter). Examples of molecules containing single bonds are shown in Figures 3.4 and 3.5. Several of these contain two sorts of electron pair; **bond pairs**, and non-bonding or **lone pairs** which are not involved in bonds but which could become so in some cases. The total number of electron pairs is important in determining the shape of a molecule. (This is discussed further near the end of this chapter.)

All of these molecules have electron density which surrounds the line joining the two atoms. Such bonds, arising from head-on overlap between orbitals, are called **sigma bonds**, symbolised σ-bonds. Other bonds arise from sideways overlap of orbitals (forming **pi bonds** symbolised π-bonds), for example two p orbitals, as shown in Figure 3.6.

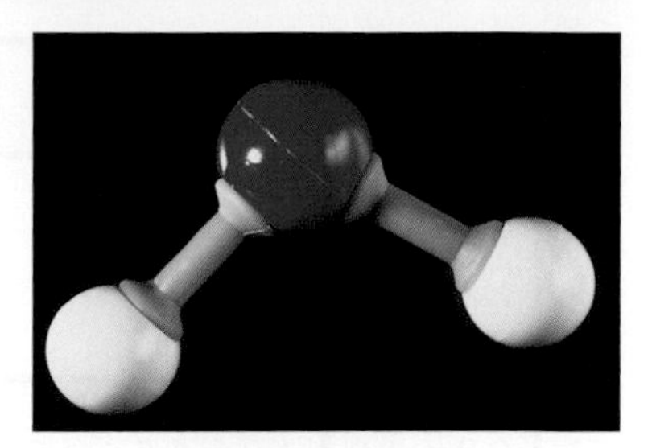

Figure 3.5 Model of a water molecule.

Figure 3.4 Single bonds, represented by dot-and-cross diagrams and by overlapping orbitals.

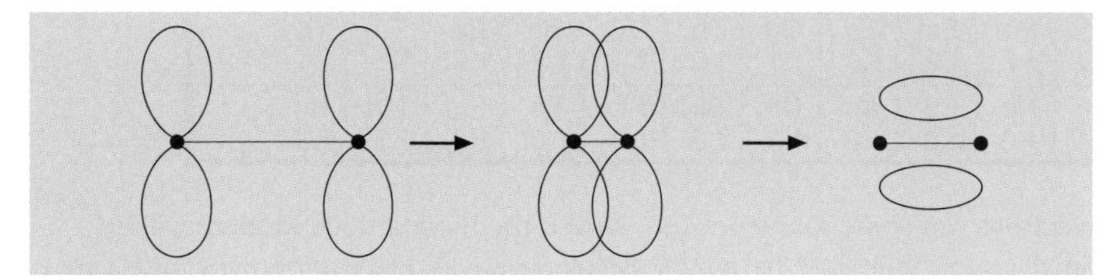

Figure 3.6 Sideways overlap of p orbitals, forming a pi bond.

These bonds arise where atoms form multiple, that is **double** or **triple bonds**. Double bonds (═) contain one σ-bond and one π-bond; triple bonds (≡) contain one σ-bond and two π-bonds. So although multiple bonds are written as though they are equivalent in molecular structure, they are not.

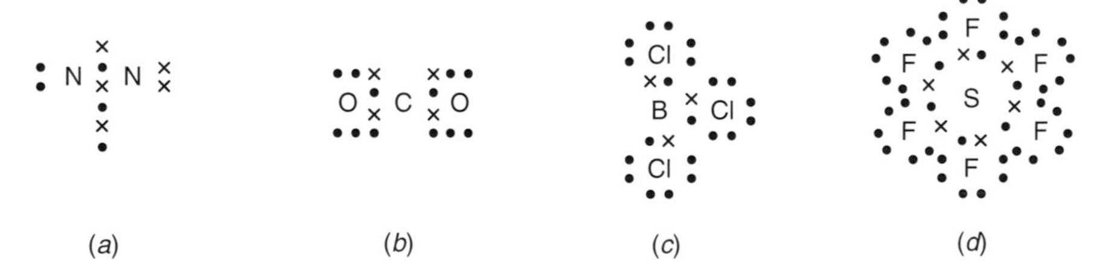

Figure 3.7 (a) and (b) Nitrogen and carbon dioxide have triple and double bonds respectively, formed by sharing more than one pair of electrons between the bonded atoms. (c) Boron trichloride is an electron-deficient molecule; the boron has only six electrons in its outer shell. (d) Sulphur hexafluoride shows 'expansion of the octet'; atoms in Period 3 and beyond can have more than eight electrons in their outer shell. In SF_6, sulphur has 12.

The nature of the chemical bond was clearly elucidated (in a book of that name) by Linus Pauling (Figure 3.8), one of the few to have won two Nobel prizes.

Polarity in bonds and in molecules

A covalent bond is polar if the electrons in the bond are unequally shared. This depends on the difference in **electronegativity** of the bonded atoms; the larger the electronegativity difference the more polar the bond.

Figure 3.8 Linus Pauling, exponent of the nature of the chemical bond and double Nobel Laureate.

Electronegativity is a measure of how strongly an atom attracts electrons when in a covalent bond. It is not the same as electron affinity, which is precisely defined for an individual atom in the gas phase (see section on Electron affinity in Chapter 1). The smaller an atom is, the closer the bonding

H 2.1	He																
Li 1.0	Be 1.5											B 2.0	C 2.5	N 3.0	O 3.5	F 4.0	Ne
Na 0.9	Mg 1.2											Al 1.5	Si 1.8	P 2.1	S 2.5	Cl 3.0	Ar
K 0.8	Ca 1.0	Sc 1.3	Ti 1.5	V 1.6	Cr 1.6	Mn 1.5	Fe 1.8	Co 1.8	Ni 1.8	Cu 1.9	Zn 1.6	Ga 1.6	Ge 1.8	As 2.0	Se 2.4	Br 2.8	Kr
Rb 0.8	Sr 1.0	Y 1.2	Zr 1.4	Nb 1.6	Mo 1.8	Tc 1.9	Ru 2.2	Rh 2.2	Pd 2.2	Ag 1.9	Cd 1.7	In 1.7	Sn 1.8	Sb 1.9	Te 2.1	I 2.5	Xe
Cs 0.7	Ba 0.9	La 1.1	Hf 1.3	Ta 1.5	W 1.7	Re 1.9	Os 2.2	Ir 2.2	Pt 2.2	Au 2.4	Hg 1.9	Tl 1.8	Pb 1.8	Bi 1.9	Po 2.0	At 2.2	Rn
Fr	Ra	Sc															

Ce 1.1	Pr 1.1	Nd 1.2	Pm 1.2	Sm 1.2	Eu 1.1	Gd 1.1	Tb 1.2	Dy 1.1	Ho 1.2	Er 1.2	Tm 1.2	Yb 1.1	Lu 1.2
Th 1.3	Pa 1.5	U 1.7	Np 1.3	Pu 1.3	Am 1.3	Cm 1.3	Bk 1.3	Cf 1.3	Es 1.3	Fm 1.3	Md 1.3	No 1.3	Lr

Figure 3.9 Pauling electronegativities of the elements. Electronegativity increases from left to right across the Periodic Table.

electrons can come to the nucleus, and so the more attraction there will be. Small atoms, therefore, have high electronegativity. Electronegativity increases from left to right across the Periodic Table, and decreases going down a group. There are various scales of electronegativity, the most commonly used being the Pauling scale in which the most electronegative element, fluorine, is arbitrarily assigned a value of 4. Pauling electronegativities are given in Figure 3.9. Electronegativity differences affect not only the polarity of the bond, but also the bond length, with larger differences giving shorter bonds than would be expected from the radii of the atoms alone.

Although a molecule may have polar bonds it does not necessarily follow that the molecule itself is polar. The dipole which the polar bond forms experiences a turning effect or torque in an electric field, and the polarity's direction is therefore important. The positive and negative ends of the dipoles are usually represented by the symbols δ^+ and δ^-. The molecule will be polar if the dipoles on each of the bonds do not cancel, so that water and ammonia are polar molecules. However CO_2, which is linear, and CCl_4, which is tetrahedral, are not polar since the dipoles from each of the bonds cancel in the molecule as a whole (Figure 3.10).

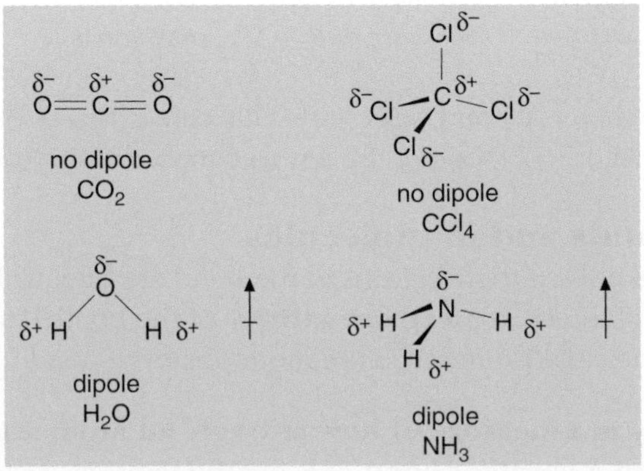

Figure 3.10 Molecules with polar bonds do not necessarily have an overall dipole moment. (The arrows show the direction of the dipole with the cross at the positive end.)

If molecules contain lone pairs, these also contribute to the overall polarity. Thus nitrogen trifluoride, NF_3, which on consideration of the bond polarities alone might be thought to be extremely polar, has very little polarity because of the cancelling effect of the lone electron pair. Its geometry is the same as that of ammonia. In this case, the lone pair more or less cancels the polarity of the N–F bonds (Figure 3.11).

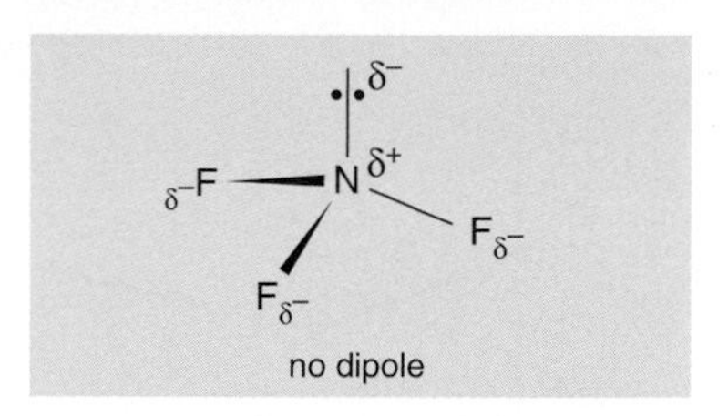

Figure 3.11 Nitrogen trifluoride has the same geometry as ammonia, but very little polarity.

Intermolecular forces

Bonds within molecules (**intramolecular bonds**) are strong and difficult to break, which is why many reactions involving covalent substances are slow. Bonds between molecules, or **intermolecular bonds**, are relatively weak, and rely on dipole–dipole attractions, which may be between permanent dipoles (the strongest) or between temporary or induced dipoles which are weaker. Temporary or induced dipole forces are generically called **van der Waals' forces**. We now consider these in more detail.

Hydrogen bonding

The strongest intermolecular force is the hydrogen bond. It is an electrostatic attraction between a strongly δ^+ hydrogen atom attached covalently to the highly electronegative elements fluorine, nitrogen or oxygen, and a strongly δ^- fluorine, nitrogen or oxygen atom on another molecule. In some cases the interaction may be within the same molecule. The strongest hydrogen bond is in hydrogen fluoride, HF, 150 kJ mol^{-1}, but many are in the range 20–60 kJ mol^{-1}. Hydrogen bonds are usually shown as broken lines, and are longer than covalent bonds (Figure 3.12). The three atoms involved are co-linear.

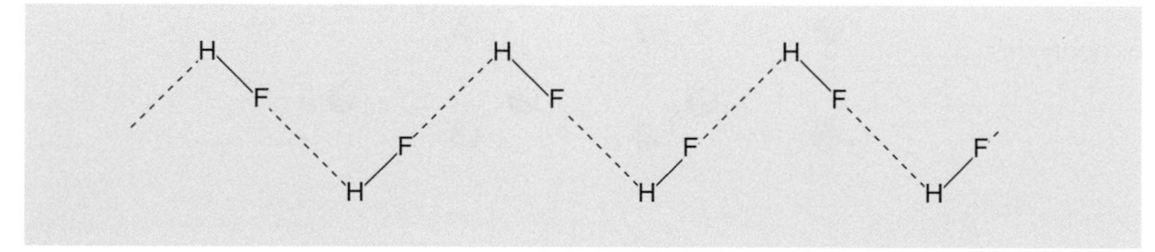

Figure 3.12 Hydrogen bonding between hydrogen fluoride molecules, giving a zig-zag chain structure.

Hydrogen bonding has a considerable effect on the properties of the substances which have it. Water has extensive hydrogen bonds in the liquid, with the result that water has a much higher boiling temperature, T_b, than would be expected by comparison with the other hydrides of Group 6. The same is true of ammonia, NH_3, compared with other Group 5 hydrides, and of hydrogen fluoride compared with other hydrides in Group 7. The effects are illustrated in Figure 3.13.

As water cools, hydrogen bonds form in greater quantity. Since they are long, they space the molecules out, so the density of the liquid decreases. At 0°C the liquid freezes to a structure where the molecules are held in rings of six, each molecule being hydrogen bonded to four others. This very ordered and open structure (Figure 3.14) has a lower density than liquid water at 0°C, so ice shows the unique property of floating in its liquid – and enables iced drinks to cool by convection and save the bother of stirring them.

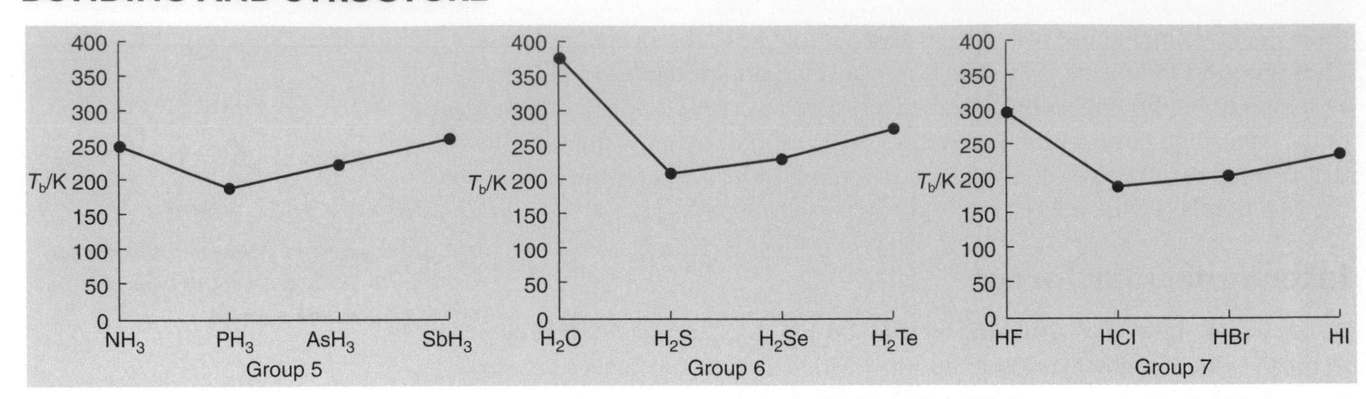

Figure 3.13 Boiling temperatures of Groups 5, 6 and 7 hydrides.

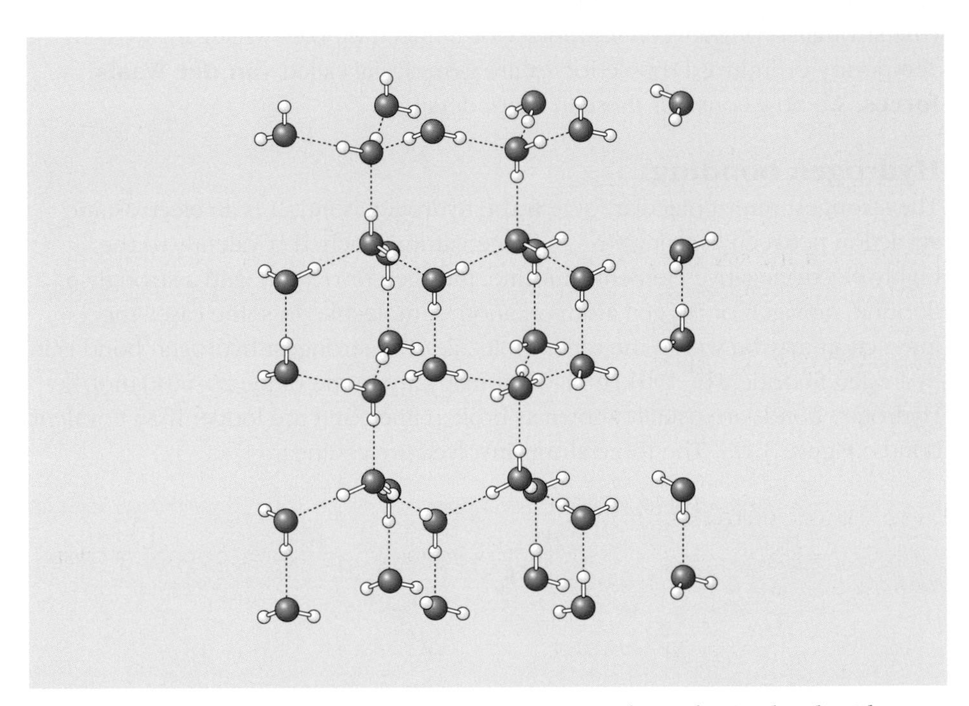

Figure 3.14 The structure of ice. Its 'open' structure gives it a lower density than liquid water.

CHO
H — C — OH
HO — C — H
H — C — OH
H — C — OH
CH_2OH
glucose

Figure 3.15 The OH groups in glucose make it very water-soluble.

Hydrogen bonds affect solubility in water. Compounds which can hydrogen bond with the water will be very soluble even if they have quite a lot of organic content. Sugars, such as glucose, which have large numbers of OH groups that hydrogen bond with water, are among the most soluble organic substances known (Figure 3.15). Benzene, C_6H_6, does not dissolve significantly in water, but phenol, C_6H_5OH, is appreciably soluble; smaller alcohols (C_1– C_3) are miscible with water in all proportions.

The structure of proteins and of nucleic acids depends strongly on hydrogen bonding. Proteins are polymers of amino acids, and long sections of the molecule coil into helices, as hydrogen bonds hold the turns of the helix together. The same is true of cellulose, a polymer of glucose. Nucleic acids, with their famous double helix structure, have the two halves of the helix attached by hydrogen bonding.

Dipole–dipole forces

Weaker forces exist between molecules which are permanently polar but which cannot form hydrogen bonds. The δ^+ and δ^- parts of the molecules attract electrostatically, and give boiling temperatures which are significantly higher than those of non-polar molecules of similar size. Thus propanone,

$$(CH_3)_2C{=}O$$

and butane, $CH_3CH_2CH_2CH_3$, have a similar size but the polar propanone boils at 329 K, while butane boils at 273 K.

Dispersion forces

Temporary dipoles form between molecules because the electron density is somewhat mobile within the molecule, and causes temporary δ^+ and δ^- areas within the molecule. A δ^+ on one molecule will induce a δ^- on a nearby one, which then causes further δ^+ to form, and so on (Figure 3.16).

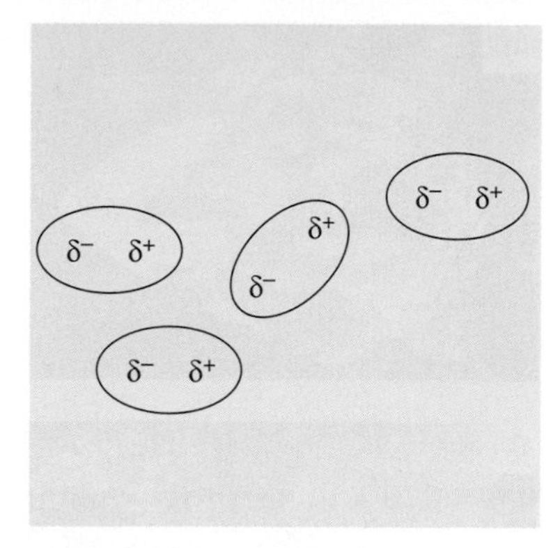

Figure 3.16 Temporary dipoles lead to a net attraction between molecules.

These dipole effects are constantly shifting, but there is a net attraction between the molecules. The larger the molecule, the more opportunities there are for these dipoles to form since there are more electrons, and the greater the intermolecular force and hence the boiling or melting temperature. The increase of T_b with increasing chain length for the alkanes, for example, is due to the increasing size of the molecules and the increased dispersion forces between the molecules (Table 3.3).

Table 3.3 *Boiling temperatures of the alkanes*

Alkane	T_b/K
CH_4	109
C_2H_6	185
C_3H_8	231
C_4H_{10}	273
C_5H_{12}	309
C_6H_{14}	342
$C_{10}H_{22}$	447
$C_{20}H_{42}$	617

Bonding and properties

The bulk properties of substances depend on the bonding. We shall consider each type in turn.

Giant molecular substances

These have covalent bonds, which extend over long distances in the solid; a crystal of visible size can consist essentially of one molecule. The best-known example is

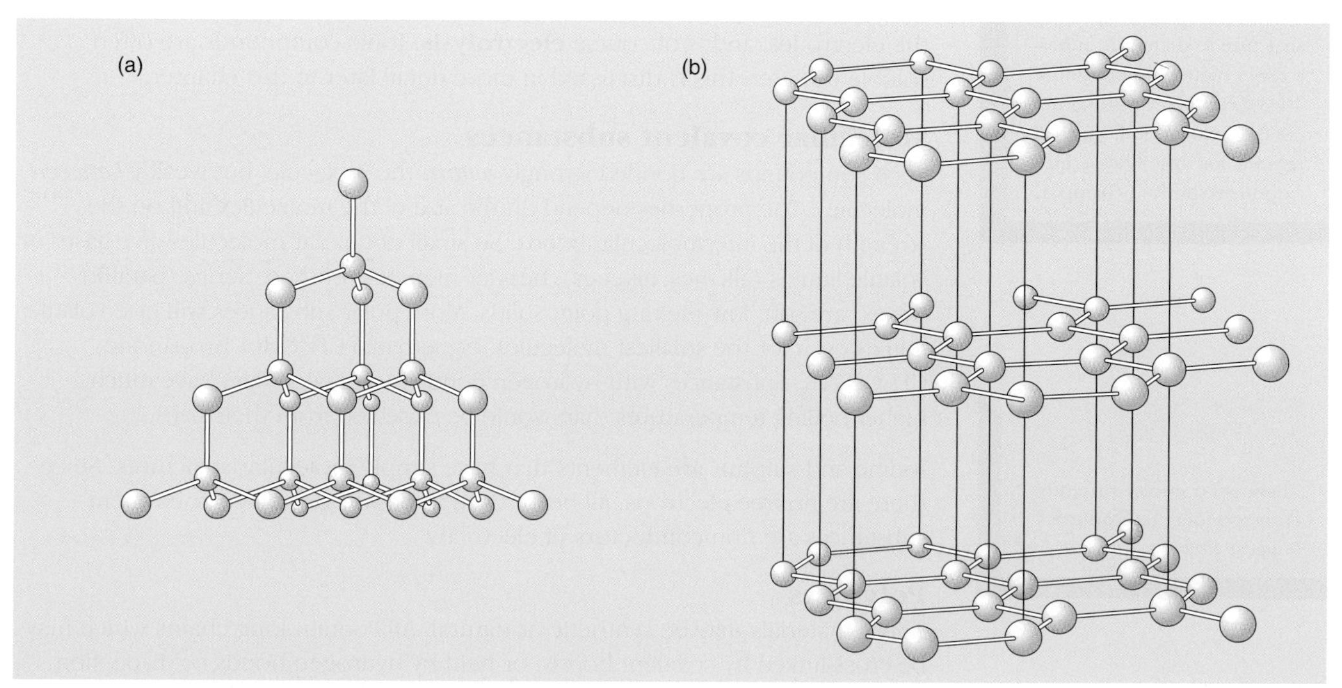

Figure 3.17 Molecular structure of (a) diamond and (b) graphite.

Figure 3.18 (a) Diamond; (b) crystalline graphite.

diamond, which has layers of chair-shaped six-membered rings of carbon joined in every direction by σ-bonds arranged tetrahedrally (Figure 3.17a). Silicon dioxide has a related structure. Giant structures have extremely high melting temperatures, for example over 3800 K for diamond, and because bonds throughout the crystal are strong the crystals are very hard. The crystal lattice is very stiff and readily transmits vibration, so diamond is a good thermal conductor, but there are no free electrons and no ions, so it is a very poor electrical conductor.

Graphite is unique. The four bonds between the carbon atoms are 3 σ-, and one π-bond forming sheets of flat hexagons (Figure 3.17b). The π-electrons are not localised between the carbon atoms; instead they are free to move along but not between the layers. Van der Waals' forces attract the layers to one another, but these are much weaker than the bonds within the layers. The crystal therefore consists of layers of giant molecules, and has a high melting temperature (over 3900 K), but the layers can slide over one another so the substance feels greasy and is used as a dry lubricant. The free electrons allow conduction of electricity along the layers only, not at right angles to them, unlike in metals where the electrons can travel in any direction. Graphite and diamond look very different (Figure 3.18).

Ionic substances

These also have strong electrostatic bonds throughout their lattice, i.e. attractions between oppositely charged ions. Thus they have high melting temperatures, which are also related to the amount of ionic character of the bonding.

They are hard but, if stressed, layers of the crystal may slide so that ions of the same charge come next to one another and repel, thus breaking the crystal. In other words, they are brittle.

The solid does not conduct electricity, since there are no charge carriers which can move, but molten ionic substances have mobile ions which will allow conduction. Such conduction is always accompanied by chemical changes at the electrodes, and is of course **electrolysis**. Ionic compounds are often soluble in water; this is discussed in more detail later in this chapter.

> **QUESTION**
>
> Silicon shows a similar structure to diamond. It has a lower melting temperature, 1683 K compared with about 3800 K for diamond. Suggest reasons for this. Silicon and carbon are both in Group 4.

Molecular covalent substances

Such compounds are bonded strongly *within* the molecule, but weakly *between* molecules. The properties depend on the size of the molecules and on the strength of the intermolecular bonds. So small non-polar molecules give gases or volatile liquids (alkanes, alkenes); heavier members of these series (paraffin waxes) are soft, low-melting-point solids. More polar substances will give volatile liquids even for the smallest molecules, e.g. ethanal CH_3CHO, propanone CH_3COCH_3. Substances with hydrogen bonds (water, alcohols) have much higher boiling temperatures than would be expected from their sizes.

Iodine and sulphur are elements that have simple molecular structures. Since there are no free electrons, all being used in bonding, molecular covalent substances are non-conductors of electricity.

> **QUESTION**
>
> There is no silicon structure corresponding to graphite. Suggest why not.

Polymers

These materials may be synthetic or natural. All contain long chains which may be cross-linked by covalent bonds, or held by hydrogen bonds or dispersion forces, or, in the case of silicates, by forces of ionic attraction between the negatively charged chains of silicate and positive metal ions. The mechanical properties of the polymer depend on the extent of cross-linking; and on whether crystallites, which are areas of crystallinity where the polymer chains are aligned, can form.

Synthetic organic polymers

Synthetic organic polymers are made from alkenes by addition reactions, or from reactions between organic molecules having two functional groups which can undergo condensation reactions (see *Transition Metals, Quantitative Kinetics and Applied Organic Chemistry*, Nelson Thornes, 2003). They are mixtures, since the chain length is variable, and so they do not show a sharp melting temperature. Instead they soften over a range of temperature which depends on their structure. Polymers that are formed by radical reactions, for example low density poly(ethene), are branched and cross-linked (bonds are formed between different chains) because the intermediates are so reactive and rather indiscriminate in their attack. There are few crystallites, and the material is flexible, elastic and translucent.

The ultimate polymer (for now), is the result of the highly controlled type of polymerisation invented by Ziegler and Natta using a catalyst complex of titanium trichloride and triethylaluminium. With this it is possible to ensure that the side chains are all on the same side of the chain, that they alternate to one side or the other, or that they are aligned at random. Because the conditions are more controlled than in radical polymerisations, there are more crystallites, and the polymers are stiffer and are opaque.

QUESTION

Suggest why polymers do not usually have a sharp melting temperature, but melt over a range of temperatures.

Natural polymers

Natural organic polymers include proteins, polysaccharides such as cellulose, starch and glycogen, and nucleic acids. All of these are produced by condensation reactions. Proteins may have structural functions (muscle, say) or catalytic ones (enzymes).

Inorganic polymers are widely distributed, for example there are the hugely abundant silicates. Phosphoric acid also forms a variety of polymeric structures on heating. The details of natural polymer chemistry are outside our present concerns.

Intermolecular forces in the liquid state

The forces between molecules in liquids are no different in type from those of solids, merely in degree. So intermolecular forces in liquids can be van der Waals, dipole–dipole or hydrogen bonds. The difference is that the particles in liquids are more energetic than those in solids, so the bonds are not particularly directional and liquids have no particular shape. When a liquid boils, the particles have to be separated from each other, and they then have to be given enough energy to break free from the liquid surface, so the boiling temperatures of liquids depend on both the *magnitude* of the interparticle forces and the *masses* of the particles themselves.

In the noble gases, the interatomic forces are very small; the boiling temperatures rise with increasing relative atomic mass (Table 3.4) and are due to increasing size.

A similar point can be made with respect to the hydrides of Group 4. All the bonds in these molecules are polar, although the tetrahedral shape means that the molecule as a whole is not polar since the polarities cancel. The boiling temperatures increase with increasing size of the molecule (Table 3.5).

Table 3.4 *The boiling temperatures of the noble gases*

Element	Relative atomic mass	T_b/K
helium	4.0	4
neon	20.2	27
argon	39.9	87
krypton	83.8	121
xenon	131.3	166

Table 3.5 *The boiling temperatures of Group 4 hydrides*

Hydride	Relative molecular mass	T_b/K
CH_4	16.0	81.6
SiH_4	32.1	161
GeH_4	76.6	185
SnH_4	122.7	221

(PbH_4 probably does not exist.)

With the hydrides of Groups 5, 6 and 7, the pattern changes dramatically. The lightest elements (N, O, F) in these groups are all electronegative enough to show hydrogen bonding as the principal intermolecular force in their hydrides. Hydrides of the heavier elements in these groups do not show the same ability, and so the boiling temperatures are much lower than those of the lightest ones (see Figure 3.13, p. 38).

Change of state

Solids have a fixed shape. The bonds between the particles in the lattice are strong enough to prevent the particles from moving large distances, although they will vibrate about a mean position in the crystal lattice. These vibrations arise because the crystal is not at absolute zero, and the amplitude of the vibrations increases with increasing temperature.

Consider what happens as a solid is heated. At some temperature the vibrations will be sufficient to overcome the forces holding the crystal together. This temperature, which is the melting temperature, T_m, will depend upon the strength of the interparticle forces. Thus a lot of hydrogen bonding produces materials of quite high melting temperature, for example glucose, whose $T_m = 423$ K; whereas molecules of similar size with van der Waals' forces only will have a much lower melting temperature, as with dodecane, whose $T_m = 264$ K.

While the compound is melting, the temperature remains constant since the heat is being used to break the interparticle forces. Once the substance is liquid, further heating increases the energy of the particles, some of which will now have enough energy to break free from the liquid surface. This is how evaporation occurs. If the temperature continues to rise, the number of molecules escaping will also rise, and so will the vapour pressure of the liquid. At the boiling temperature T_b the vapour pressure of the liquid is the same as the external pressure, and bubbles of vapour are produced throughout the liquid, and it is boiling. The heat being put in to the liquid is being used to overcome the interparticle forces so that the molecules can escape the liquid, and so the temperature remains constant.

In the case of atomic or molecular substances, the vaporised particles are usually the same molecules or atoms that are present in the liquid. This is not always the case; the vapour of sodium contains Na_2, for example, and that from sulphur contains S_2 and S_4. In the case of ionic compounds, the vapours usually consist of ion pairs, sodium chloride vapour containing Na^+Cl^- pairs.

QUESTION

Suggest reasons why sodium chloride vapour is regarded as a collection of ion pairs rather than as NaCl molecules.

The shapes of molecules and ions

The great German chemist Hermann Kolbe wrote in 1877 of the 'fantasies' of J.H. van't Hoff (Figure 3.19), who had postulated that molecules have definite shapes. Kolbe did not spare the feelings of his target in an amazingly vitriolic article which made reference to 'flights of fancy on a Pegasus from the veterinary college', van't Hoff being employed at the time at the veterinary college of Utrecht; but van't Hoff was eventually vindicated. He received the first Nobel Prize in chemistry; Kolbe had by then been dead some 16 years.

Figure 3.19 J.H. van't Hoff, winner of the first Nobel Prize in chemistry.

That molecules and polyatomic ions do have shapes is well known. There are two rules which enable us to predict these shapes:

1. The shape adopted is the one that puts the electron pairs in the valence shell, i.e the outer shell (whether bonding or lone pairs), as far apart as possible, since this minimises repulsion between the negative charges and clouds.
2. The order of repulsion between the various sorts of electron pairs is

   bp – bp *less than* bp – lp *less than* lp – lp

   where 'bp' means bonding or bond pair and 'lp' means lone pair.

Rule 2 does not affect the shape in a fundamental way, but does modify the bond angles somewhat. Consider some examples.

Two bond pairs

This arrangement is found in beryllium chloride vapour, $BeCl_2$, which is linear (the solid consists of polymeric chains). The bonds are as far apart as possible at 180°:

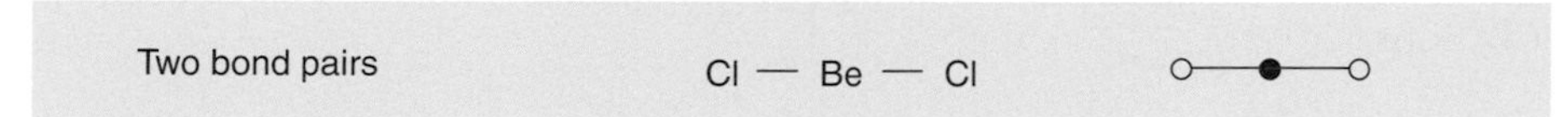

Three bond pairs

This produces a trigonal planar structure, where the molecule is flat and the bond angles are 120°. Boron trifluoride, BF_3, is a good example:

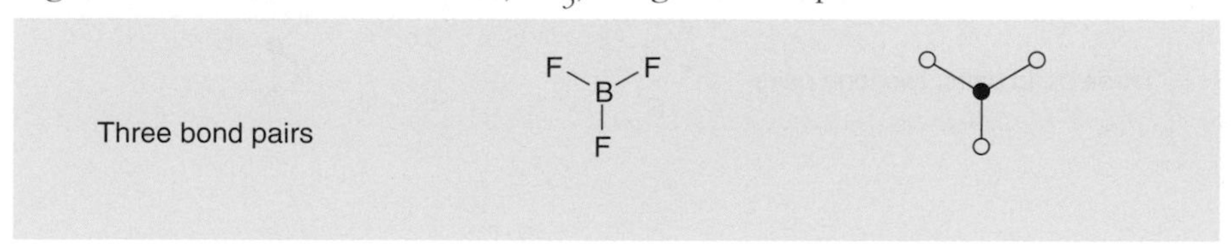

Four bond pairs

The fundamental arrangement is tetrahedral, e.g. as in methane (and all other saturated carbon atoms), with a bond angle of 109.5°:

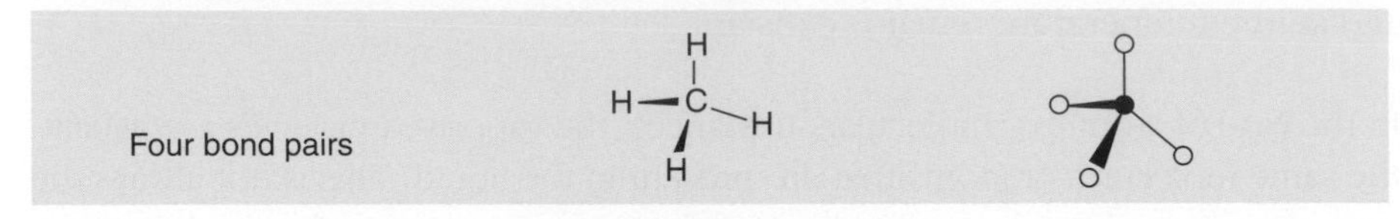

Three bond pairs, one lone pair

The arrangement of orbitals is tetrahedral, but since only three of the pairs are bonding pairs the molecule is called 'pyramidal', e.g. ammonia. The repulsion between the lone pair and the bond pairs is more than that between the bond pairs because the lone pair is pulled towards the nucleus and is rather fatter than the other orbitals. The H–N–H bond angle is therefore compressed slightly to 107° from the tetrahedral angle of 109° 28':

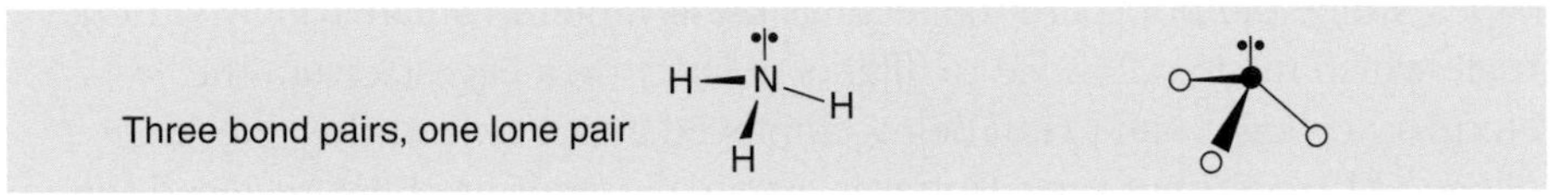

On forming an ammonium ion, the electrons all become bond pairs, and NH_4^+ is a regular tetrahedron. There is no difference between the dative bond and the other three once formed.

Two bond pairs, two lone pairs

This arrangement is found in water, whose molecule is bent as shown. Since there are two lone pairs the bond angle is compressed even more, to around 104°. On forming H_3O^+ a pyramidal structure results, and in extremely acidic solution H_4O^{2+} forms, in small quantities, and is tetrahedral:

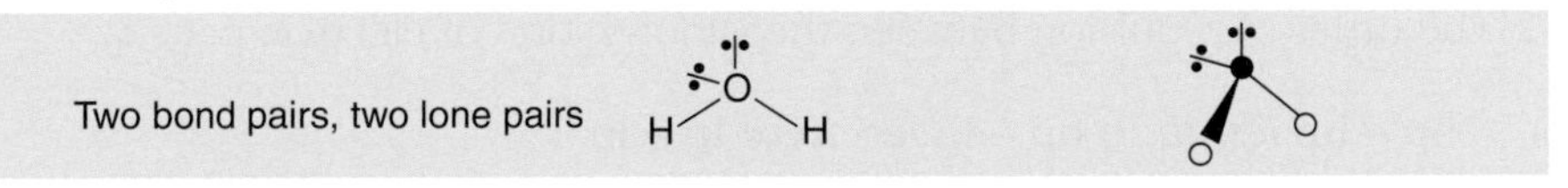

Five bond pairs

This is exemplified by gaseous PCl_5. Three bonds are at 120° in a plane, with two others at 90° to this plane, giving the trigonal bipyramid shape:

Five bond pairs

Cl
Cl
P—Cl
Cl
Cl

Solid PCl_5 has a different structure, with tetrahedral PCl_4^+ and octahedral PCl_6^- ions.

Three bond pairs, two lone pairs

Where lone pairs are present among five electron pairs, the lone pairs always occupy the equatorial plane. Thus ClF_3 is T-shaped, with two lone pairs:

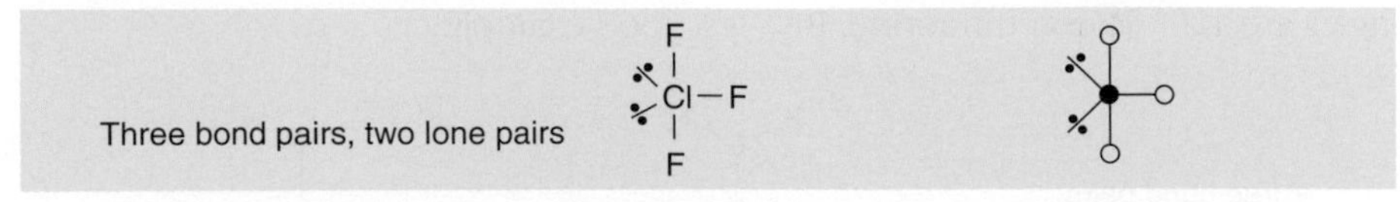

Six bond pairs

All the bond angles are 90°, and the structure is a regular octahedron. This is found in SF_6, and in the huge number of hexacoordinated transition metal complexes, such as hexaquairon(II), $[Fe(H_2O)_6]^{2+}$:

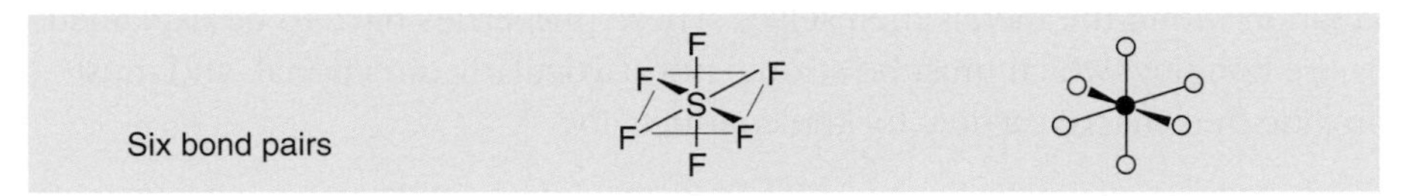

Therefore all that needs to be known in order to predict shapes is the number of electron pairs that have to be accommodated, and it does not matter whether neutral molecules or ions are being considered. What has not been done here is to explain the mechanism by which the orbitals achieve these shapes; that lies outside our present concerns.

Molecules and ions with multiple bonds

Stereochemically, multiple bonds behave as if they were single bonds. Figure 3.20 shows some examples of molecules and ions that have such bonds, and the shapes that result from the ideas of electron pair repulsion mentioned above. Many of these species have more than eight electrons in the outer shell of the central atom, and so do not 'obey' the octet 'rule'.

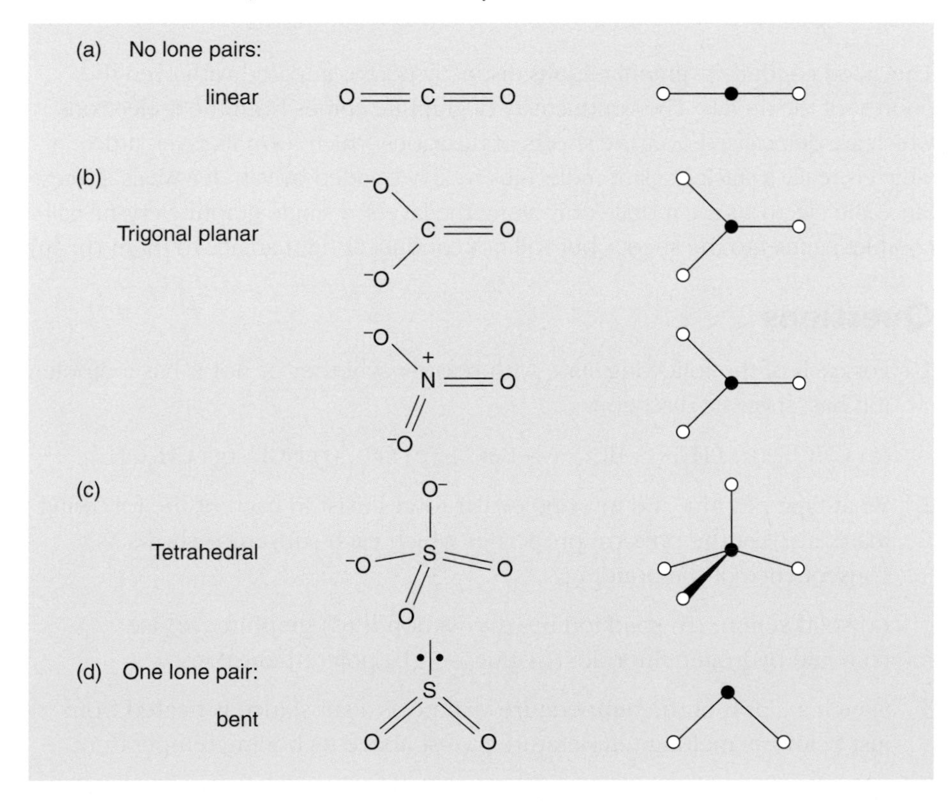

Figure 3.20 Shapes of molecules and ions with multiple bonds.

Metallic bonding

The principal physical characteristics of metals are their plastic nature (that is, they can be deformed without breaking), and their electrical conductivity, which is much greater than most non-metals, but which shows quite wide variation within the metals themselves. These properties have to be explained by the bonding, which must be strong, not particularly directional, and must provide free charge carriers for the conductivity.

The outer electrons in metals are not paired up into individual bonds, but rather are free to move around the whole lattice structure of the metal which is formed from the positive ions. The free movement of this 'sea' of electrons allows for electrical conductivity, and the non-directional nature of the attractions between the ions and the electrons allows malleability. A good macroscopic analogy is that of a group of ball-bearings, representing the ions, covered in grease, representing the sea of electrons.

Table 3.6 *Resistances relative to silver of selected elements*

Element	Relative resistance
Ag	1.0
Cu	1.05
Fe	6.2
Pb	13.5
Ni	4.3
Cr	7.9
Hg	65
Au	1.5
C	887
Si	10^{18}
S	$\sim 10^{23}$

The best conductor is silver, followed by copper; other metals are much poorer, and resistance wires, e.g. for electric fires, are made from alloys such as Ni/Cr (nichrome). Table 3.6 gives some values for the resistance of selected elements, relative to silver which is the best conductor there is, other than superconductors.

The 'good conductor' graphite looks distinctly poor compared with even the poorest of the metals. The conductivity of graphite comes from the π-electrons which are delocalised over the sheets of hexagons which form its layer lattice, which is really a stack of giant molecules weakly bonded by van der Waals' forces. Since the electrons are mobile only along the layers, a single graphite crystal will conduct parallel to the sheets, but will not conduct at right angles to them (p. 40).

Questions

1 For each of the following state with reasons whether or not it has a dipole; if it has, show its direction.

(a) $CHCl_3$; (b) PH_3; (c) BCl_3; (d) $BeCl_2$; (e) CS_2; (f) NCl_3; (g) CH_3OH.

2 What type of intra- and intermolecular forces exist in each of the following materials? List the physical properties which each substance shows as a consequence of the bonding:

(a) solid xenon; (b) solid iodine; (c) diamond; (d) graphite; (e) ice; (f) liquid hydrogen fluoride; (g) glucose; (h) poly(ethene).

3 Sketch a graph of the temperature vs time as a substance is heated from just below its melting temperature to just above its boiling temperature.

4 Draw the following so as to show their shape. Give reasons for your choice.

(a) BCl_3; (b) PCl_3; (c) $BeCl_2$; (d) CO_2; (e) NH_3; (f) $SnCl_2$; (g) $SnCl_4$; (h) PCl_5; (i) SF_6; (j) PCl_6^-; (k) NO_2^-; (l) NO_2^+; (m) CO_3^{2-}; (n) SO_4^{2-}; (o) SO_2; (p) SO_3^{2-}; (q) BrF_3; (r) XeF_4; (s) BrF_4^-.

The Periodic Table I

Mendeleev's original Periodic Table of 1869 was not the first attempt to classify the elements. Earlier efforts by Döbereiner and Lothar Meyer were important to Mendeleev's work. He arranged the majority of the elements by their atomic mass (or weight, as it was called then). Having no notion of atomic properties, this was all that was available to him at the time. This arrangement led to some anomalies, for example with tellurium and iodine. Iodine has relatively fewer neutrons than tellurium, so the mass of a tellurium nucleus is larger than that for iodine. This would have put Te in Group 7 and iodine in Group 6. This is absurd in terms of chemical similarity, so Mendeleev, confident in the predictive powers of his table and unaware of the existence of the neutron or any other subatomic particle, ignored the problem in the belief that it would go away. When it became clear early this century that the arrangement of the elements in the Periodic Table depends on electronic structure and hence on the atomic number rather than the atomic mass, the problem did indeed go away. See p. 84 for a version of the Periodic Table commonly used today.

The number of groups in each block of the Periodic Table depends on the subshell that is being filled. Thus the s-block (Groups 1 and 2) has s electrons as the valence (outer) shell electrons; there are two groups because the s orbital can hold only two electrons. Similarly, the p-block has six groups in it since the p subshell contains up to six electrons. The d-block has 10 elements in each row, the five d orbitals containing at most 10 electrons.

Some physical properties of the elements of Period 3, Na to Ar

Melting and boiling temperature

The melting and boiling temperatures of the elements of Period 3 are shown in Table 4.1. The pattern of increasing melting and boiling temperatures to a maximum at aluminium and silicon reflects the binding energy of the solids and the liquids. This in turn depends on the extent to which orbitals on adjacent atoms can overlap in the case of the metallic lattices, or the extent of dispersion forces (and hence the size of the molecule) in the case of the molecular covalent elements. The more orbitals there are that can be involved in the bonding from the valence shell, the tighter the binding; this is highest for silicon, which can use all four valence-shell orbitals. The number is fewer

Table 4.1 *Melting and boiling temperatures for the elements of Period 3*

Element	Na	Mg	Al	Si	P	S	Cl	Ar
T_m/K	371	922	933	1683	317	392	172	84
T_b/K	1156	1380	2740	2328	553	718	238	87
Structure		metallic		giant covalent	P_4	S_8	Cl_2	atoms

for atoms earlier in the period where atoms have too few electrons to utilise all their orbitals, and later where orbitals contain non-bonding electron pairs.

These elements show four bond types: metallic (sodium, magnesium, aluminium) but with no crystal structure in common; giant covalent (silicon); molecular covalent (phosphorus, sulphur and chlorine) but with no molecular structure in common; and van der Waals' forces between atoms (argon).

Sodium to aluminium

The melting temperature of a metal depends upon three things: the size of the atom, the number of electrons that are used in bonding the lattice, and the crystal structure. The increase in melting temperature from Na to Al parallels other quantities that depend on the strength of the metal lattice that is the enthalpy of atomisation ΔH_a, the enthalpy of vaporisation ΔH_{vap}, and the enthalpy of fusion ΔH_{fus}. These quantities are given in Table 4.2.

BACKGROUND

ΔH_a: the enthalpy of atomisation is the heat change accompanying conversion of an element into one mole of its atoms.
ΔH_{vap}: the enthalpy of vaporisation is the heat change accompanying the conversion of one mole of a substance into its vapour at the same temperature.
ΔH_{fus}: the enthalpy of fusion is the heat change accompanying the conversion of one mole of solid substance into its liquid at the same temperature.
See also Topic 2.1 (Unit Test 2).

Table 4.2 *Enthalpy changes related to binding energy for the metals in Period 3*

	ΔH_a/kJ mol^{-1}	ΔH_{vap}/kJ mol^{-1}	ΔH_{fus}/kJ mol^{-1}
Sodium	107	89	2.60
Magnesium	148	129	8.95
Aluminium	326	294	10.7

The model used for metallic bonding is familiar enough, where metal ions are packed into and attracted to a 'sea' of electrons that are delocalised over the whole metal lattice. The electron sea comes from the valence (outer shell) electrons of the atom. This model, the Drude–Lorentz theory, describes well many things about metals (malleability, ductility, density), but cannot account for the very wide differences in electrical conductivity.

As the atomic number increases across the period, the atoms get smaller. This is because the electrons are being added to the same shell, but the nuclear charge, and hence the attraction for the electrons, is increasing. Since the atoms are smaller, they can pack more closely. Closer packing means better orbital overlap, and a higher melting temperature. Furthermore, sodium has only one valence shell electron to delocalise around the metal lattice, whereas magnesium has two and aluminium has three.

The three metals do not have a crystal structure in common. Although the details of crystal structures are outside our present concerns, it is worth knowing that sodium (and the alkali metals generally) adopts a cubic packing that does not bring the atoms as close as is possible. Each sodium atom has eight nearest neighbours. Magnesium and aluminium, by contrast, are close-packed structures; although they do not pack in quite the same way, each atom has 12 nearest neighbours.

The combination of these factors leads to a sharp rise in melting temperature between sodium and magnesium, with only a small change between magnesium and aluminium.

Boiling temperatures are often less sensitive to changes in structure than are melting temperatures. However, at the boiling temperature, each of the metals has a liquid structure which has been described as a blurred version of the crystal. Thus liquid sodium at 100°C has atoms with eight nearest neighbours, as in the crystal; aluminium has 10–11 compared with 12 in the crystal. The much higher boiling temperature of Al compared with Na results from the increased number of binding electrons and the smaller size of the atoms, just as in the solid. The enthalpy changes also increase, for the same reason.

The three metals all give monatomic vapours, even just above the boiling temperature.

Silicon

Silicon has a giant covalent structure, like diamond but with longer and therefore weaker bonds. Its very high melting temperature reflects the necessity to break covalent intramolecular bonds if the element is to melt. Silicon can use four orbitals to bond, so it has the highest binding energy of all the elements in Period 3, as well as the highest enthalpy of vaporisation, $377\,kJ\,mol^{-1}$. Boiling requires that the covalent bonds should break, which they do completely since silicon vapour is also monatomic.

Phosphorus, sulphur and chlorine

All are molecular covalent, but the molecules have no common features. The melting points are related to the size of the molecule; the larger it is, the greater the opportunity for van der Waals' (dispersion) forces to occur, and therefore sulphur, with the largest molecule (Table 4.1), has the highest melting and boiling temperatures.

- Phosphorus has three solid forms: white, red and black. The white form is fairly volatile, which is why it is toxic. It smells of garlic. The solid contains P_4 molecules that are tetrahedral and which have relatively weak intermolecular van der Waals' forces; the melting temperature is therefore low.
- Sulphur also has two solid structures, both having S_8 molecules with the atoms in a ring. The forces between the rings are van der Waals' forces, but because the molecule is larger than that of phosphorus the melting temperature is higher. The melting temperature quoted is that for monoclinic sulphur, which is the stable form between 369 K and the melting temperature.
- Chlorine has the smallest of the molecules, Cl_2. The van der Waals' forces are therefore the weakest of these three elements, and it has the lowest melting temperature of the three as well.
- Argon is monatomic, and the only forces present are the very small van der Waals' forces between small atoms, so the melting temperature is very low.

QUESTION

Explain the nature of the intermolecular forces in chlorine.

Electrical conductivity

The electrical conductivity of each of the solid elements in Period 3 is given in Table 4.3, relative to silver.

Table 4.3 *The conductivity of solid Period 3 elements, relative to Ag = 1.0*

Na	Mg	Al	Si	P	S
0.35	0.36	0.61	10^{-18}	10^{-17}	10^{-23}

The ability to conduct electricity results from mobile charge carriers, which can be electrons or ions. The three metals conduct electricity because the delocalised electrons in the metal lattice will drift in the presence of an electric field. The Drude–Lorentz theory does not account for the wide differences in metallic conductivity, although other models of the metallic structure do.

None of the other elements possesses any mobile electrons, so these elements are all non-conductors.

Silicon has no structure analogous to that of graphite. The Si atom is larger than the C atom, and this means that the p-overlap needed to form multiple bonds is inefficient in silicon. The Si=Si bonds formed would be much weaker than two single Si–Si bonds. This is not true for one C=C bond compared with two C–C bonds.

Ionisation energy

The ionisation energies of the atoms in Period 3 mirror those of the elements in Period 2 (lithium to neon), and for the same reasons. This was discussed earlier in Chapter 1, pp. 10–12. The values in Period 3 are lower than those in Period 2, because the atoms are larger and so the outer electrons are screened from the nuclear charge more effectively by the inner-shell electrons.

Questions

1 Plot on a (small) graph the first ionisation energies of the elements from sodium to argon, and account for the shape obtained.

2 Use data from a data book to plot a graph of atomic radius vs atomic number for the elements of Periods 2 and 3 (Li to Ar). Account for the difference in the graphs between Groups 2 and 3.

3 Sketch the structures of:
(a) the giant covalent lattice of silicon
(b) the molecule P_4
(c) the molecule S_8.

4 Silicon has no compounds in which the silicon atom forms double bonds with other elements. Phosphorus, by contrast, does form double bonds with other elements. Suggest why silicon and phosphorus are different in this respect.

5 Explain how the Drude–Lorentz theory of metallic bonding accounts for the following properties of metals:
(a) their malleability
(b) their density
(c) their electrical conductivity.

Oxidation/reduction: an introduction

Introduction

Oxidation and reduction are found with all but four elements in the Periodic Table, not just with the transition metals, although they show these reactions to such an extent that they could be accused of self-indulgence.

When magnesium reacts with oxygen (Figure 5.1)

$$2Mg(s) + O_2(g) \rightarrow 2MgO(s)$$

the product contains Mg^{2+} and O^{2-} ions. Reaction with oxygen is pretty clearly oxidation. The reaction of magnesium with chlorine

$$Mg(s) + Cl_2(g) \rightarrow MgCl_2(s)$$

gives a compound with Mg^{2+} and Cl^- ions. In both cases the magnesium atom has lost electrons, so as far as the magnesium is concerned the reactions are the same. This idea is generalised into the definition of oxidation as loss of electrons. Reduction is therefore the gain of electrons. Since electrons don't vanish from the universe, oxidation and reduction occur together in **redox** reactions.

DEFINITION

Oxidation is electron loss.
Reduction is electron gain.

MNEMONIC

OIL RIG:

oxidation **is l**oss

reduction **is g**ain

Oxidation numbers

For simple monatomic ions such as Fe^{2+} it's easy to see when they are oxidised (to Fe^{3+}) or reduced (to Fe). For ions such as NO_3^- or SO_3^{2-} which also undergo oxidation and reduction it is not always so easy to see what is happening in terms of electrons. To assist this, the idea of **oxidation number** or **oxidation state** is used. The two terms are usually used interchangeably, so that an atom may have a particular oxidation number or be in a particular oxidation state.

Figure 5.1 The use of magnesium flares in photography being demonstrated at an early meeting of the British Association in Birmingham (1865).

Each element in a compound is treated as though it is an ion, no matter what the actual nature of the bonding. If during a reaction the 'charge' on the 'ion' becomes more positive, then that part of the compound has been oxidised. The inverted commas are used because the compound may not be ionic; it is taken to be so for this electronic book-keeping exercise.

Some atoms have defined oxidation states. There are three simple rules to start with:

1. Elements have an oxidation number of zero.
2. A simple monatomic ion has an oxidation number that is the same as its charge. The oxidation number is given in Arabic numerals with the appropriate sign, except when naming compounds, where Roman numerals are used. Thus Fe^{3+} is iron(+3), but a compound of it would be, say, iron(III) chloride.
3. The oxidation number of hydrogen is (+1) in most of its compounds, and that of oxygen is (–2). (There are some exceptions to this which will be considered once the main rules are established.) Fluorine is always (–1).

DEFINITION

The **Oxidation number** or **oxidation state** of an atom is the difference between the number of electrons associated with an element bonded in a compound and the element itself.

To see how to use oxidation numbers, consider the reaction between chlorine and bromide ions. The oxidation numbers are shown underneath each substance:

$$Cl_2 + 2Br^- \rightarrow 2Cl^- + Br_2$$

$$(0) \quad (-1) \quad (-1) \quad (0)$$

The chlorine has been reduced because its oxidation number has decreased; the bromide ion has been oxidised because its oxidation number has increased.

That example is simple enough to make the use of oxidation numbers unnecessary in such a formal way. However, consider the reaction of manganate(VII) ions with iron(II) ions:

$$MnO_4^-(aq) + 5Fe^{2+}(aq) + 8H^+(aq) \rightarrow Mn^{2+}(aq) + 5Fe^{3+}(aq) + 4H_2O(l)$$

It is easy enough to see that the iron(II) ions have been oxidised to iron(III), but not so easy to see why there are five of them for every MnO_4^-. If we use the oxidation number (–2) for oxygen, pretending that the manganate (VII) ion is wholly ionically bonded we get:

$$[Mn^{x+}(O^{2-})_4]^-$$

from which $x = 7$. Thus since the product is Mn^{2+}, there must have been five electrons added to the manganese in MnO_4^- to reduce it. These come from five Fe^{2+} ions, losing one electron each. The manganese in MnO_4^- is not actually a (7+) ion (indeed you will know from Chapter 3 that such a highly charged ion would be so polarising as to give covalent bonding) but it behaves in redox reactions as though it is.

Another common oxidising agent is dichromate(VI), $Cr_2O_7^{2-}$, which is not quite as powerful as manganate(VII). Using the same idea as before, in this case $[(Cr^{y+})_2(O^{2-})_7]^{2-}$, it is not hard to get $y = 6$. When dichromate(VI) is reduced, it forms two Cr^{3+} ions, a six-electron change.

To see which part of a compound is negative and which is positive in finding oxidation numbers the electronegativity is used. The more electronegative atom has the negative oxidation number. So in ammonia, NH_3, nitrogen is the more electronegative and has oxidation number (–3); in nitrite, NO_2^- where oxygen is more electronegative it is (+3), and in nitrate, NO_3^-, (+5). Carbon in carbon dioxide, CO_2 is (+4), but in methane, CH_4, it is (–4) . This is because carbon is more electronegative than hydrogen. The oxidation number is not the same as an element's valency or combining power. Carbon shows valency four only.

Sometimes hydrogen does not have oxidation number +1. In sodium hydride, NaH, it is combined with a less electronegative atom, and so hydrogen has the oxidation number (–1) in ionic hydrides. Oxygen shows positive oxidation numbers only when combined with fluorine, e.g. it is (+2) in oxygen difluoride, F_2O.

Compounds or ions which apparently show fractional oxidation numbers usually have atoms of the same type with two or more different oxidation numbers. In tri-iron tetroxide, Fe_3O_4, for example, there is one Fe(+2) and two Fe(+3), the oxide behaving as $FeO.Fe_2O_3$. In trilead tetroxide, Pb_3O_4, the compound behaves as $2PbO.PbO_2$, i.e. Pb(+2) and Pb(+4). There are some exceptions: in the superoxide ion O_2^-, oxygen has an oxidation number of $-\frac{1}{2}$; in the tetrathionate ion, $S_4O_6^{2-}$, sulphur has an oxidation state of $+2\frac{1}{2}$.

Oxidation numbers and redox reactions

The usefulness of oxidation numbers lies largely in the derivation of equations for redox reactions. These always involve something being oxidised and something being reduced, and so can be split into two half-reactions, one for the oxidation and one for the reduction. These are the processes that occur at the electrodes if the reaction is done in an electrochemical cell (see *Transition Metals, Quantitative Kinetics and Applied Organic Chemistry,* Nelson Thornes, 2003, Chapter 1). Knowing a few half-reactions means that you can combine them into a wide variety of full reactions since a given oxidising agent is, under given conditions, usually reduced in the same way whatever the reducing agent.

The oxidation of bromide ions by chlorine, mentioned earlier, affords a simple example. Chlorine is reduced to chloride:

$$Cl_2 + 2e^- \rightarrow 2Cl^-$$

and bromide is oxidised to bromine:

$$2Br^- \rightarrow Br_2 + 2e^-$$

Addition of these two half-reactions gives the full reaction, since the two electrons given up by the bromide ions are the *same* electrons that are gained by the chlorine atoms.

$$Cl_2 + 2Br^- \rightarrow Br_2 + 2Cl^-$$

When half-reactions are combined, the number of electrons in each must be the same, which may require one or both half-reactions to be multiplied by an integer. Note that a half-reaction cannot be written unless the starting and finishing substances are known; there is no general rule for deriving the product from the reagent.

Manganate(VII) ion in acid solution is reduced to Mn^{2+}:

$$MnO_4^- \rightarrow Mn^{2+}$$

but other things are clearly needed. Five electrons are required for the reduction from manganese(+7) to manganese(+2):

$$MnO_4^- + 5e^- \rightarrow Mn^{2+},$$

but there are still the four oxygens in the ion to deal with. Remembering that these are regarded as O^{2-}, they are converted to water with the hydrogen ions from the acid:

$$MnO_4^- + 5e^- + 8H^+ \rightarrow Mn^{2+} + 4H_2O$$

This is the half-reaction for manganate(VII) ion reduction in acid.

The half-reaction for oxidation of iron(II) to iron(III) is:

$$Fe^{2+} \rightarrow Fe^{3+} + e^-$$

and since five electrons are needed to reduce the manganate(VII) this equation is multiplied by 5 and then added to the MnO_4^- half-reaction to ensure that the same number of electrons are gained by MnO_4^- as are lost by Fe^{2+}. We get:

$$MnO_4^- + 5Fe^{2+} + 8H^+ \rightarrow Mn^{2+} + 5Fe^{3+} + 4H_2O$$

In general if O^{2-} needs to be disposed of from the left-hand side, H^+ will do it. With dichromate(VI),

$$Cr_2O_7^{2-} + 6e^- + 14H^+ \rightarrow 2Cr^{3+} + 7H_2O$$

	$6e^-$		$14H^+$
	reducing $2 \times Cr(+6)$ to $2 \times Cr(+3)$		for reaction with $7O^{2-}$

In many cases reducing agents require more oxygen, for instance when NO_2^- becomes NO_3^-. The need is for O^{2-}, which can be obtained from water, $2H^+$ remaining:

$$NO_2^- + H_2O \rightarrow NO_3^- + 2e^- + 2H^+$$

the $2e^-$ coming from the oxidation of N(+3) to N(+5). The electrons and hydrogen ions do not combine to form hydrogen, since the electrons aren't free but will have been given to the oxidising agent.

Suppose that the oxidising agent is potassium manganate(VII) acidified with H_2SO_4. Writing the two half-reactions,

$$MnO_4^- + 8H^+ + 5e^- \rightarrow Mn^{2+} + 4H_2O$$

$$NO_2^- + H_2O \rightarrow NO_3^- + 2e^- + 2H^+$$

the electrons donated are made the same as the number of electrons received if we multiply the top half-reaction by 2 and the bottom by 5,

$$2MnO_4^- + 16H^+ + 10e^- \rightarrow 2Mn^{2+} + 8H_2O$$

$$5NO_2^- + 5H_2O \rightarrow 5NO_3^- + 10e^- + 10H^+$$

We then add to get the overall reaction

$$2MnO_4^- + 5NO_2^- + 6H^+ \rightarrow 2Mn^{2+} + 5NO_3^- + 3H_2O$$

Consider one more example, the reaction of dichromate(VI) ion with hydrogen peroxide in acid solution. Hydrogen peroxide can behave both as a reducing agent, as here, or as an oxidising agent.

The half-reactions are

$$Cr_2O_7^{2-} + 6e^- + 14H^+ \rightarrow 2Cr^{3+} + 7H_2O$$

$$H_2O_2 \rightarrow O_2 + 2H^+ + 2e^-$$

Multiplying the second equation by 3 and adding gives the overall reaction

$$Cr_2O_7^{2-} + 3H_2O_2 + 8H^+ \rightarrow 2Cr^{3+} + 3O_2 + 7H_2O$$

Using half-reactions, large numbers of redox reactions can be derived which would otherwise have to be learnt individually. Table 5.1 gives some which, apart from those involving halogens, need not be learnt for AS.

You should not pass over any redox reaction without writing out the half-reactions and being clear what the changes in the oxidation numbers are.

Table 5.1 *Some half-reactions*

Half-reaction
$MnO_4^- + 8H^+ + 5e^- \rightarrow Mn^{2+} + 4H_2O$
$Cr_2O_7^{2-} + 14H^+ + 6e^- \rightarrow 2Cr^{3+} + 7H_2O$
$H_2O_2 \rightarrow O_2 + 2H^+ + 2e^-$ (reducing agent)
$H_2O_2 + 2H^+ + 2e^- \rightarrow 2H_2O$ (oxidising agent)
$NO_2^- + H_2O \rightarrow NO_3^- + 2H^+ + 2e^-$
$SO_3^{2-} + H_2O \rightarrow SO_4^{2-} + 2H^+ + 2e^-$
$2ClO_3^- + 12H^+ + 10e^- \rightarrow Cl_2 + 6H_2O$
$FeO_4^{2-} + 8H^+ + 3e^- \rightarrow Fe^{3+} + 4H_2O$
$2IO_3^- + 12H^+ + 10e^- \rightarrow I_2 + 6H_2O$
$MnO_2 + 4H^+ + 2e^- \rightarrow Mn^{2+} + 2H_2O$
$PbO_2 + 4H^+ + 2e^- \rightarrow Pb^{2+} + 2H_2O$
$2S_2O_3^{2-} \rightarrow S_4O_6^{2-} + 2e^-$
$VO^{2+} + 2H^+ + e^- \rightarrow V^{3+} + H_2O$
$VO_2^+ + 2H^+ + e^- \rightarrow VO^{2+} + H_2O$
$I_2 + 2e^- \rightarrow 2I^-$
$ClO_3^- + 6H^+ + 5e^- \rightarrow \frac{1}{2}Cl_2 + 3H_2O$

Strength of oxidising agents

Oxidising agents differ in their ability to remove electrons. This is covered in more detail in Unit 5 of the A2 course, but is relevant to the extraction of bromine from seawater. Chlorine is a strong enough oxidising agent to remove electrons from bromide ions to give bromine, but iodine is a weaker oxidising agent and will not oxidise bromide ions.

Questions

1 Find the oxidation numbers of:
(a) Fe in $FeCl_3$
(b) Cl in NaCl
(c) S in SO_4^{2-}
(d) S in SO_3^{2-}
(e) Mn in MnO_4^{2-}
(f) Mn in MnO_2
(g) C in CO
(h) C in CO_2
(i) C in CCl_4
(j) Cr in CrO_3
(k) Os in OsO_4
(l) Br in BrF_3
(m) Cl in HOCl
(n) Cl in $HClO_3$
(o) Cl in $HClO_4$
(p) Fe in $[Fe(CN)_6]^{4-}$
(q) Cu in $CuCl_4^{2-}$
(r) O in F_2O

2 Combine the required half-reactions from Table 5.1 and elsewhere in this chapter to obtain the oxidation–reduction reactions between:
(a) MnO_4^- and H_2O_2
(b) MnO_4^- and SO_3^{2-}
(c) MnO_4^- and NO_2^-
(d) $Cr_2O_7^{2-}$ and H_2O_2
(e) MnO_4^- and I^-
(f) PbO_2 and Cl^-
(g) V^{3+} and MnO_4^- (two reactions successively)
(h) $S_2O_3^{2-}$ and I_2
(i) ClO_3^- and Cl^-

3 Separate the following oxidation–reduction reactions into their two constituent half-reactions:
(a) $2FeCl_2 + Cl_2 \rightarrow 2FeCl_3$
(b) $2KBr + Cl_2 \rightarrow 2KCl + Br_2$
(c) $MnO_2 + 4HCl \rightarrow MnCl_2 + Cl_2 + 2H_2O$
(d) $MnO_4^- + 5Fe^{2+} + 8H^+ \rightarrow Mn^{2+} + 5Fe^{3+} + 4H_2O$
(e) $2MnO_4^- + 5C_2O_4^{2-} + 16H^+ \rightarrow 2Mn^{2+} + 5CO_2 + 8H_2O$
(f) $2S_2O_3^{2-} + I_2 \rightarrow S_4O_6^{2-} + 2I^-$
(g) $H_2SO_4 + 8HI \rightarrow H_2S + 4I_2 + 4H_2O$
(h) $IO_3^- + 5I^- + 6H^+ \rightarrow 3I_2 + 3H_2O$
(i) $2Cu^{2+} + 4I^- \rightarrow 2CuI + I_2$
(j) $Zn + Cu^{2+} \rightarrow Zn^{2+} + Cu$
(k) $Fe_2O_3 + 2Al \rightarrow Al_2O_3 + 2Fe$
(m) $KIO_3 + 2Na_2SO_3 \rightarrow KIO + 2Na_2SO_4$
(n) $Sn + 4HNO_3 \rightarrow SnO_2 + 4NO_2 + 2H_2O$
(o) $Cl_2 + 2NaOH \rightarrow NaCl + NaOCl + H_2O$
(p) $2H_2O_2 \rightarrow 2H_2O + O_2$

6 The Periodic Table: Groups 1 and 2

The s-block elements

Groups 1 and 2 of the Periodic Table are called the s-block elements since the outer electrons are in s orbitals. Group 1 elements are also called the alkali metals and Group 2 elements the alkaline earth metals. The two groups have fairly simple chemistries with clear trends, and are metals which are very reactive and of very low density when compared with metals as a whole. The compounds of both groups are almost wholly ionic, the Group 1 and 2 metals having oxidation numbers of +1 and +2 respectively. They have no other oxidation states. The compounds are usually colourless unless a transition metal is present in the anion; for example potassium manganate(VII), $KMnO_4$, is purple owing to the purple MnO_4^- ion. The superoxides are exceptions.

The inorganic chemistry of metals typically deals with their reactions with oxygen and water, since these are abundant naturally, and with chlorine, a strong oxidising agent.

Group 1

Compared with the metals of the d-block, the alkali metals are not very dense, and have low melting and boiling temperatures; these properties are given in Table 6.1. The metals are soft, and can easily be cut with a knife, potassium being roughly the texture of plasticine. The abundance of each element in the Earth's crust is given in parts per million; only sodium and potassium are common.

Table 6.1 *Some properties of the Group 1 elements*

Element	Atomic radius/pm	Ionic radius/pm	Density/ $g\,cm^{-3}$	Melting temp/°C	Boiling temp/°C	Abundance/ ppm
lithium	133	60	0.53	181	1330	65
sodium	157	95	0.97	98	890	28300
potassium	203	133	0.86	63	774	25900
rubidium	216	148	1.53	39	688	310
caesium	235	169	1.88	29	690	7

All of these properties come from the relatively weak metallic bonding; there is only one electron available per atom to be delocalised around the crystal, and the atoms themselves are the largest of their period. The crystal structure of the metals is body-centred cubic (Figure 6.1a), a type of packing which does not bring the atoms as close together as the commoner hexagonal close-packed or face-centred cubic lattices can (Figure 6.1b). Lithium is the least dense of all solid elements, and lithium, sodium and potassium float on water – as they react with it. As the atoms get larger, the bonding becomes weaker, so the melting temperatures and the hardness decrease with increasing atomic number. The density rises, however, because the mass of the atom increases more rapidly than its size with increasing atomic number.

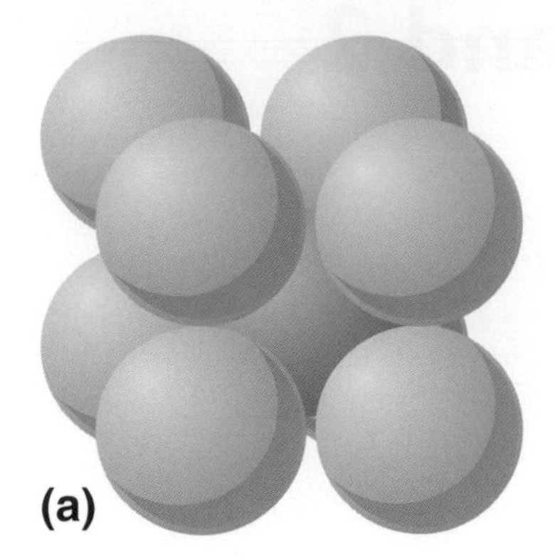

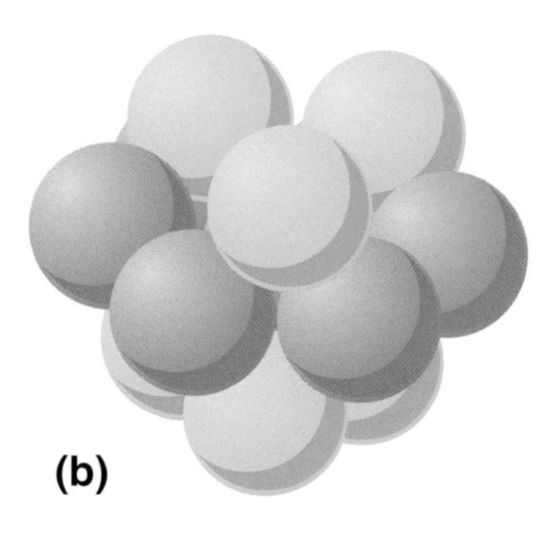

Figure 6.1(a) Body-centred cubic arrangement of atoms; (b) hexagonal close-packed arrangement.

The first two ionisation energies for the alkali metals are given in Table 6.2, below.

Table 6.2 *Ionisation energies of Group 1 elements*

Element	First ionisation energy/kJ mol^{-1}	Second ionisation energy/kJ mol^{-1}
lithium	520	7298
sodium	496	4563
potassium	419	3051
rubidium	403	2632
caesium	376	2420

The alkali metals have one s electron outside a closed inert-gas configuration, from $1s^22s^1$ for lithium to $[Xe]6s^1$ for caesium. The first ionization energies decrease with increasing relative atomic mass. The nuclear charge increases on descending the group, but the increasing number of electron shells between the nucleus and the outermost electron shield the outer electron from the nuclear charge. An alternative view is that the inner electron shells repel the outer electron. Either way, the increase in shielding outweighs the increase in nuclear charge so that the attraction for the electron falls, and so does the first ionisation energy.

Second ionisation energies for a given atom are considerably larger than the first, since to remove the second electron would require breaking into the inert-gas structure. This penultimate shell is much closer to the nucleus (look at Table 6.1 to see the difference between the size of the atom and the size of the ion), the nuclear attraction is much larger, and so the electrons are more strongly held.

Group 2

Various physical quantities concerning Group 2, the alkaline earth metals, are given in Table 6.3 below.

Table 6.3 *Some physical quantities for Group 2 elements*

Element	Atomic radius/pm	Ionic radius/pm	Density/ g cm^{-3}	Melting temp/°C	Boiling temp/°C	Abundance/ ppm
beryllium	89	31	1.85	1278	2477	6
magnesium	136	65	1.74	649	1110	20900
calcium	174	97	1.54	839	1487	36300
strontium	191	113	2.6	769	1380	150
barium	198	135	3.51	725	1640	430

The trends in atomic radius and ionisation energy are the same on descending this group as they are for Group 1, and the explanations for this are similar. Densities are higher than Group 1 but still low for metals in general; this is partly because the atoms are smaller in Group 2, and also because the crystal structures are closely packed (apart from that of barium, which has body-centred cubic packing), and two electrons per atom are available for metallic bonding. Melting points are therefore higher than the corresponding metals in Group 1, and the metals are harder.

Table 6.4 lists the first three ionisation energies for the alkaline earth metals.

Table 6.4 *Ionisation energies for Group 2 elements*

Element	First ionisation energy/kJ mol^{-1}	Second ionisation energy/kJ mol^{-1}	Third ionisation energy/kJ mol^{-1}
beryllium	900	1757	14800
magnesium	738	1451	7740
calcium	590	1145	4940
strontium	550	1064	4120
barium	503	965	3390

The first ionisation energies are higher for Group 2 than for Group 1; the nuclear charge is greater, but the shielding of the outer electron by inner shells of electrons is similar. Thus comparing sodium and magnesium, the outer electron in each case is in the 3s shell and they have the same inner electron configuration $1s^2\ 2s^2\ 2p^6$. The nuclear charge of magnesium is +12 (from 12 protons) compared to that of sodium which is +11 (from 11 protons), and the atom is smaller. Hence the attraction for the outer electrons in magnesium is larger so they are more difficult to remove. The large jump to the third ionisation energy is because this would involve breaking into closed shells.

Some reactions of the s-block metals

The reactions of the s-block are fairly simple with few complicating features, so they can be considered together. Those of beryllium excepted, the compounds are essentially ionic, and the only oxidation states shown are the group states, +1 and +2. In both cases this is because going above these entails an ionisation energy input which could not be recouped from the lattice enthalpy of the resulting solid. Similarly, $MgCl_2$ is formed rather than MgCl because the increased lattice enthalpy more than compensates for the second ionisation energy of magnesium (see Chapter 3).

DEFINITION

A **basic** oxide is one that will react with an acid to form a salt. Most metal oxides are basic.

An **amphoteric** oxide is one that will react with either an acid or an alkali to form a salt. The oxides of beryllium, chromium, aluminium, tin and lead are amphoteric.

Reactions with oxygen

The reactions with oxygen highlight the only significant idiosyncrasies within the chemistry that concerns us, since the alkali metals do not all behave in the same way. All of the normal oxides are basic with the exception of beryllium oxide, which is amphoteric.

Lithium burns in oxygen with a deep red flame to form the usual oxide containing the O^{2-} ion:

$$4Li(s) + O_2(g) \rightarrow 2Li_2O(s)$$

Lithium oxide is a white solid.

Sodium burns in oxygen with a yellow flame to give a mixture of sodium oxide Na_2O, and sodium peroxide, Na_2O_2, the latter predominating in excess oxygen:

$$2Na(s) + O_2(g) \rightarrow Na_2O_2(s)$$

Sodium peroxide contains the peroxide ion, O_2^{2-}, in which the oxidation state of oxygen is (–1). Sodium peroxide is a pale yellow solid.

DEFINITION

Radicals are atoms or ions with a lone (single) electron. They tend to be very reactive.

The remaining alkali metals form superoxides, which contain the O_2^- ion. This has an unpaired electron and is therefore a **radical anion**; one of the oxygen atoms does not have a full octet of electrons. Superoxides are exceptional among Group 1 compounds not containing transition metal ions in that they are coloured. KO_2 is yellow, RbO_2 orange, and CsO_2 red. Potassium, for example, burns with a lilac flame to give potassium superoxide, KO_2:

$$K(s) + O_2(g) \rightarrow KO_2(s)$$

All Group 1 metals will also react readily with the oxygen in the air at room temperature. Silvery when freshly cut, they rapidly tarnish and become dull due to a coating of oxide. For this reason, lithium, sodium and potassium are usually stored under paraffin oil, while rubidium and caesium which are much more reactive are stored in sealed containers. Lithium also reacts significantly with the nitrogen in the air, and forms a mixture of lithium oxide and lithium nitride, Li_3N.

QUESTION

Give the equation representing the reaction of lithium with nitrogen. What ions are present in lithium nitride?

Except for barium, the elements of Group 2 all form their oxide on heating in oxygen. Magnesium burns with a brilliant white flame which is used for flares and in fireworks, calcium has a brick-red flame, and strontium deep red. These are the same as the colours which their compounds impart to the Bunsen flame and which are used in analysis. The equation representing the reaction of magnesium with oxygen is typical:

$$2Mg(s) + O_2(g) \rightarrow 2MgO(s)$$

The oxides are white solids, and are basic. Barium burns with an apple green flame to give a mixture of the oxide, BaO, and white barium peroxide, Ba_2O_2.

QUESTION

Give the equation for the reaction of magnesium with steam, and suggest why magnesium oxide is formed under such conditions rather than magnesium hydroxide.

In its reaction with oxygen, lithium behaves more like a Group 2 metal than a Group 1 metal. Lithium shows anomalies in other properties, and overall behaves in many ways like magnesium. This is an example of a diagonal relationship; another is that between beryllium and aluminium, which behave similarly. Beryllium oxide is amphoteric, for example, as is aluminium oxide.

Reactions with water

All the s-block metals will react readily with cold water except for beryllium and magnesium. Beryllium is not attacked at all; magnesium reacts very slowly with cold water, but much more quickly with hot water or steam. The rate of reaction is generally faster for Group 1 than Group 2 and increases on descending each group. Hydrogen is liberated, and except for magnesium where with steam magnesium oxide is formed, the hydroxide of the metal is produced. In the case of lithium, sodium and potassium, the metals float on the surface of the water, melt, and often burst into flame. The reaction of sodium with water is typical for Group 1:

$$2Na(s) \quad + \quad 2H_2O(l) \quad \rightarrow \quad 2NaOH(aq) \quad + \quad H_2(g)$$

The sodium rushes about on the surface of the water and the hydrogen liberated often burns with a golden-yellow flame. Rubidium and caesium react explosively but the reaction is essentially the same.

The only significant difference for Group 2 metals is that the hydroxide formed is much less soluble and, in the case of calcium hydroxide, precipitates to give a milky-white suspension. The metals do not float.

$$Ca(s) \quad + \quad 2H_2O(l) \quad \rightarrow \quad Ca(OH)_2(s) \quad + \quad H_2(g)$$

QUESTION

Give the equations representing the reactions occurring when carbon dioxide is passed through limewater until there is no further change.

Reactions with chlorine

All s-block elements react directly on heating with chlorine gas to form the chloride. The chlorides are ionic except for that of beryllium, which is covalent. It forms a polymer based on a linear molecule with dative covalent bonds:

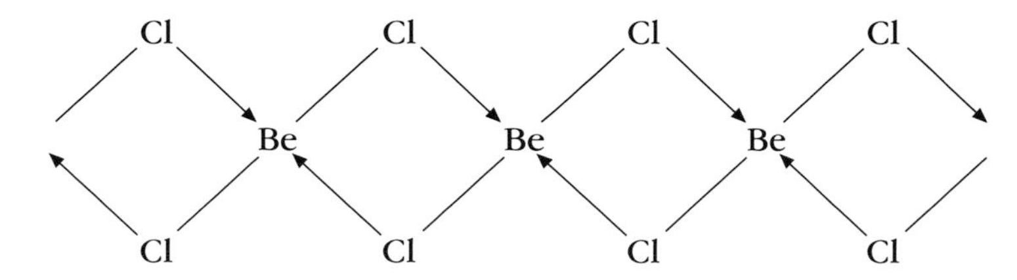

Figure 6.2 The linear molecule of solid beryllium chloride.

The reactions of sodium and magnesium with chlorine are typical:

$$2Na(s) \quad + \quad Cl_2(g) \quad \rightarrow \quad 2\ NaCl(s)$$

$$Mg(s) \quad + \quad Cl_2(g) \quad \rightarrow \quad MgCl_2(s)$$

The ionic chlorides are all soluble in water. Magnesium chloride has some covalent character, as shown by the difference between the Born–Haber cycle value for the lattice energy and the calculated value. If hydrated magnesium chloride is heated, some hydrolysis occurs and hydrogen chloride is evolved:

$$MgCl_2(s) \quad + \quad H_2O(l) \quad \rightleftharpoons \quad Mg(OH)Cl(s) \quad + \quad HCl(aq)$$

Reactions of the *s*-block oxides with water

All of the oxygen compounds mentioned earlier react with water. The oxides, containing O^{2-} ions, react to form the metal hydroxide:

$$O^{2-}(s) \quad + \quad H_2O(l) \quad \rightarrow \quad 2OH^-(aq)$$

Peroxides, containing the O_2^{2-} ion, react with water liberating hydrogen peroxide, H_2O_2:

$$O_2^{2-}(s) + 2H_2O(l) \rightarrow 2OH^-(aq) + H_2O_2(aq)$$

Superoxides, with the O_2^- ion, react to form hydrogen peroxide and oxygen

$$2O_2^- + 2H_2O(l) \rightarrow 2OH^-(aq) + H_2O_2(aq) + O_2(g)$$

Thus all the solutions will be alkaline owing to the formation of the OH^- ion in solution. The reactions are of the anion, so they are independent of the cation present and are the same for Groups 1 and 2.

Reactions of the s-block oxides with dilute acid

The reactions of the s-block oxides with acids are similar to those with water, except that a solution of the appropriate metal salt is also formed. Thus with hydrochloric acid:

$$Li_2O(s) + 2HCl(aq) \rightarrow 2LiCl(aq) + H_2O(l)$$

$$Na_2O_2(s) + 2HCl(aq) \rightarrow 2NaCl(aq) + H_2O_2(aq)$$

$$2KO_2(s) + 2HCl(aq) \rightarrow 2KCl(aq) + H_2O_2(aq) + O_2(g)$$

The solubilities of the hydroxides and sulphates of Group 2

The solubilities of the hydroxides and sulphates of Group 2 are summarised in Table 6.5.

Table 6.5 *Solubilities of Group 2 sulphates and hydroxides*

Ion	Cation radius /pm	Sulphate solubility/ mol per 100 g water	Hydroxide solubility/ mol per 100 g water
Mg^{2+}	65	1.83×10^{-1}	2.00×10^{-5}
Ca^{2+}	99	4.66×10^{-3}	1.53×10^{-3}
Sr^{2+}	113	7.11×10^{-5}	3.37×10^{-3}
Ba^{2+}	135	9.43×10^{-7}	1.50×10^{-2}

Trends in thermal stabilities of carbonates and nitrates of Groups 1 and 2

The carbonates of Group 2, and Li_2CO_3 in Group 1 decompose on heating, and clear trends are evident. The decomposition of calcium carbonate is typical:

$$CaCO_3(s) \rightarrow CaO(s) + CO_2(g)$$

This reaction is important in cement manufacture and in the extraction of iron.

QUESTION

Write the equation for the thermal decomposition of lithium carbonate, Li_2CO_3.

One view of the reasons for thermal decomposition is based on the polarising power of the cation. The larger the metal ion, the lower its polarising power because its charge density is less; small, highly charged ions are the most polarising. This causes distortion of the anion, so that oxygen atoms on the anion are pulled towards the metal ion and the bond between this oxygen atom and the rest of the anion is weakened. Salts of large polarisable anions, e.g. nitrate, carbonate, will be most stable with large, relatively non-polarising cations.

The nitrates illustrate these points further. Group 2 nitrates all decompose to the metal oxide, brown nitrogen dioxide, and oxygen. Calcium nitrate is typical:

$$2Ca(NO_3)_2(s) \rightarrow 2CaO(s) + 4NO_2(g) + O_2(g)$$

The relatively small cations form stronger lattices with oxide than with the larger nitrate. The decomposition becomes more difficult as the cation size increases, barium nitrate requiring red heat.

In the case of Group 1, the larger cations result in reactions where the product is the nitrite. This is a smaller anion than a nitrate, and so gives a higher lattice energy than nitrates do.

$$2NaNO_3(s) \rightarrow 2NaNO_2(s) + O_2(g)$$

No brown gas is evolved. The nitrates are all white and the nitrites are pale yellow.

This might seem like a lot of chemistry; however, comparisons in chemistry, and their explanation, require much time and effort.

BACKGROUND

Another version uses the idea of lattice energies. The thermal stability of a carbonate will depend on the stability of the carbonate lattice compared with the oxide lattice at the same temperature. As the cation size changes, the lattice energies of the carbonates and those of the oxides change by different factors. The lattice energies of the carbonates change little, since the carbonate ion dominates in size. However, the oxide lattice energy falls faster as the cation size increases and the sum of the radii of the ions increases. The thermal stabilities decrease in the order

$$BaCO_3 \gg SrCO_3 > CaCO_3 > MgCO_3 \gg BeCO_3$$

for Group 2. Barium carbonate does not decompose at bunsen temperatures; all other Group 2 carbonates decompose similarly to $CaCO_3$, as in the equation given above.

Group 1 cations are larger than those of Group 2, and because of the smaller cation charge the lattice energies of the carbonates are smaller too. The result of this is that the differences in lattice energies between the carbonate and the oxide are not sufficient to allow decomposition of the carbonates at normal Bunsen temperatures. The exception is lithium carbonate, which has the smallest cation; its carbonate does decompose on heating.

Flame colours and their analytical uses

Some of the s-block metals or their compounds impart a characteristic colour to a Bunsen flame: Table 6.6.

Table 6.6 *Flame colours of the s-block elements*

Group 1	lithium	carmine red
	sodium	yellow
	potassium	lilac
Group 2	calcium	brick red
	strontium	crimson red
	barium	apple green

The flame test can be used to identify the presence of these elements; the reds are difficult to describe, and in the case of Li and Sr difficult to tell apart. The flame test is performed by picking up a speck of the test compound on a clean platinum wire which has been moistened with concentrated hydrochloric acid (the chlorides being the most volatile compounds), then placing the sample at the edge of a roaring Bunsen flame (see Figure 6.3). The heat energy of the flame causes electrons to be excited within the metal atoms. When these return to lower energy levels they emit light of characteristic frequencies and therefore colour. This can be observed qualitatively, or can be analysed in a (visible) spectrometer. This shows a spectrum which consists of a series of lines of definite frequency, each line corresponding to a particular electron transition, and not equally distributed throughout the spectrum. The series of lines produced is a line spectrum and is unique to a particular element. The spectrum for sodium is shown in Figure 6.4 (p. 64). Alkali metals and alkaline earth metals have strong lines in the visible region of the spectrum. Sticks of sodium chloride were at one time sold to enable the production of sodium light in the laboratory, for example for the investigation of optical activity. Sodium light is virtually monochromatic.

Flame spectra can also be used quantitatively; the flame photometer can measure the concentrations of sodium and potassium in body fluids, for example. In astronomy, the spectra can be used to analyse the atmosphere of other planets. Helium was found on the sun, via the absorption spectrum of sunlight, before it was found on Earth.

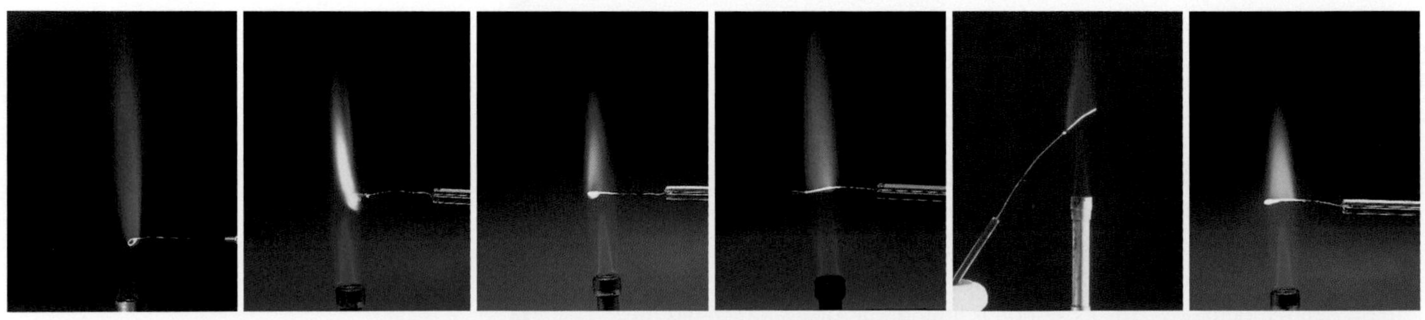

Figure 6.3 The flame colours of (left to right): lithium, sodium, potassium, calcium, strontium and barium.

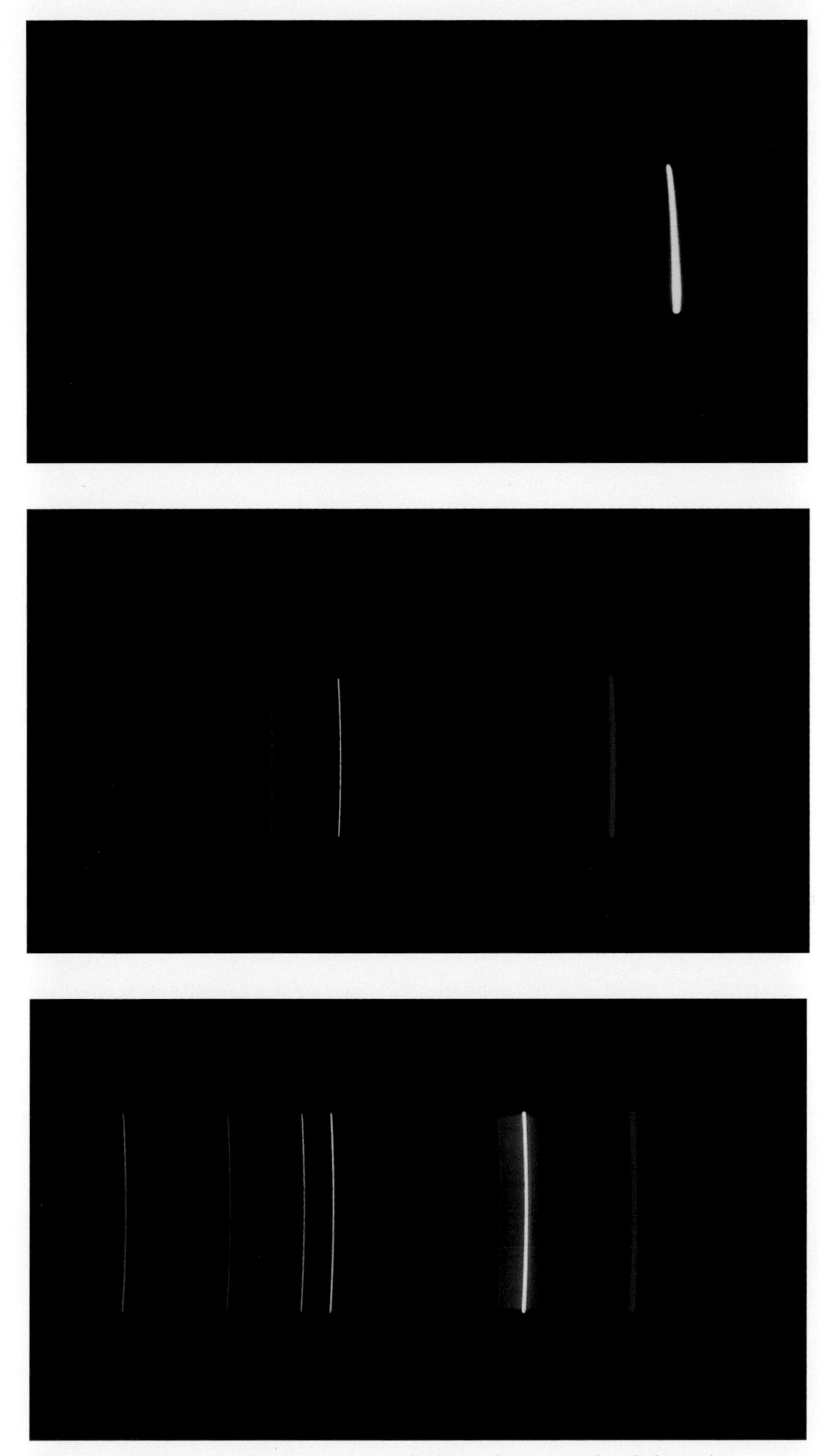

Figure 6.4 The spectrum of sodium, top (an alkali metal), compared with those of hydrogen, centre, and helium.

Questions

1 Draw (small) graphs of the data in:

(*a*) Table 6.1 vs atomic number

(*b*) Table 6.2 vs atomic number

(*c*) Table 6.3 vs atomic number

(*d*) Table 6.4 vs atomic number.

2 Lithium and magnesium ions have similar charge densities, and show a 'diagonal relationship', i.e. their chemistries are in some ways similar. Thus $LiNO_3$ decomposes in a similar way on heating to the Group 2 nitrates. Write an equation for its decomposition.

7

The halogens

The elements

The word 'halogen' means salt-former, used because of the large number of ionic, salt-like compounds which Group 7 elements form. Chlorine is one of the most significant of industrial chemicals, produced on an enormous scale.

It is used for water treatment, for production of polymers (PVC), as bleach (and in the manufacture of bleach), in making hydrochloric acid, and in many smaller-scale synthetic applications.

Figure 7.1 Chlorine.

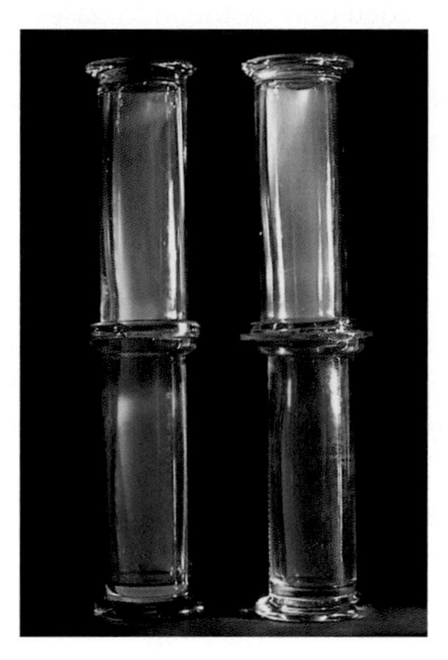

Figure 7.2 Bromine.

Figure 7.3 Iodine.

Physical properties

The principal physical properties of the halogens are given in Table 7.1.

Table 7.1 *Principal physical properties of the halogens*

Element	State at room temperature	Melting temperature/°C	Boiling temperature/°C	Atomic radius/pm	Ionic radius/pm
fluorine	pale yellow gas	−220	−188	72	136
chlorine	greenish gas	−101	−34.7	99	181
bromine	brown volatile liquid	−7.2	58.8	114	195
iodine	dark grey lustrous solid	114	184	133	216

The forces between the diatomic molecules X_2 of the halogens are van der Waals' forces, so the melting and boiling temperatures increase with increasing size of the molecules. The atomic radius increases with increasing atomic mass and hence increasing number of shells; in every case the size of the halide ion is greater than that of the parent atom since repulsions increase with the addition of the extra electron to the outer shell.

DEFINITION

The **bond dissociation enthalpy** is the enthalpy change associated with the breaking of one mole of a particular bond.

Table 7.2 *Some further properties of the halogens*

Element	1st ionisation energy/ kJ mol^{-1}	Electron affinity/ kJ mol^{-1}	Bond dissociation enthalpy/ kJ mol^{-1}
fluorine	1680	–348	158
chlorine	1260	–364	242
bromine	1140	–342	193
iodine	1010	–314	151

The ionisation energies (Table 7.2) show the expected trend with increasing size of the atom, and are the second highest of all atoms within the same period since the atoms are the second smallest; in both respects they are exceeded only by the inert gases. Knowledge of the ionisation energies is less useful than for metals. Halogens in positive oxidation states do not form positive ions as such, but are bonded covalently to more electronegative atoms, as in the oxyanions OCl^- and ClO_3^-, for example.

The electron affinities show a peak at chlorine. Since fluorine is the smallest atom with the fewest electrons, it might be thought that it would attract electrons most strongly of all. The electron is being added to an atom of small volume, however, and the repulsions from other electrons are therefore disproportionately large. The remaining electron affinities fall with increasing size of the atom, as might be expected with the accompanying fall in nuclear attraction.

The bond dissociation enthalpies show a peak at chlorine; the usual rule is that the shorter a bond is, the stronger it is, but it is possible to have too much of a good thing and in the case of fluorine the non-bonding electrons are brought so close together that they repel, and weaken the bond compared with what might have been expected. This feature is part of the reason for fluorine's very reactive nature, the high lattice energies or bond enthalpies in the compounds produced being the other part of the picture.

Tests for the halogens

There are various tests which can be used to identify chlorine, bromine and iodine, but there is no test common to all three.

TEST

Chlorine

- Bleaches litmus paper
- Oxidises Br^- to Br_2
- Oxidises I^- to I_2
- Turns starch-iodide paper blue-black

Chlorine: The tests for chlorine employ its strongly oxidising character.

Moist litmus paper is bleached rapidly; the colour is a pigment from a lichen, and chlorine oxidises it to colourless products. If blue litmus is used, it very briefly turns red before being bleached.

Chlorine will oxidise bromides to bromine, and iodides to iodine. In both cases the solutions turn brown, though that of iodine is much darker, and if excess chlorine is used solid iodine will precipitate. The first of these reactions

$$Cl_2(aq) + 2Br^-(aq) \rightarrow Br_2(aq) + 2Cl^-(aq)$$

is used to obtain bromine from seawater, so isn't just of analytical interest. If an immiscible organic solvent is added to the reaction mixtures, for example hexane, the organic layer will be coloured brown with bromine, or purple with iodine.

Moist starch-iodide paper (paper impregnated with starch and potassium iodide) turns blue-black with chlorine. The iodide ions in the paper are oxidised to iodine, and the starch then reacts with this to give a blue-black compound.

Bromine: Bromine will bleach litmus paper, but does much more slowly than chlorine.

Bromine liberates iodine from iodides in aqueous solution; the solution darkens, and an immiscible organic solvent can be used to show the presence of iodine, turning purple.

Bromine turns moist starch-iodide paper blue-black, for the same reasons as chlorine does.

Bromine will react with paper moistened with the dye fluorescein to turn it scarlet. The product is eosin, an ingredient of red ink.

TEST

Bromine
- Bleaches litmus paper but slowly
- Liberates iodine from iodide solutions
- Turns starch-iodide paper blue-black
- Turns fluorescein scarlet

Iodine: Iodine is without effect on litmus.

Iodine will turn moist starch-iodide paper blue-black, but will also do the same to starch solution alone, which neither chlorine nor bromine will do.

Iodine is purple in solution in organic solvents which have no oxygen, for example hexane or methylbenzene. The presence of oxygen in a solvent gives a browner cast to the solution, so the antiseptic Tincture of Iodine, which is a solution in ethanol, is brown. This difference in colour can be used as a qualitative test for oxygen-containing solvents in organic chemistry.

TEST

Iodine
- No effect on litmus paper
- Turns starch solution blue-black
- Is purple in organic solvents without oxygen

The hydrogen halides

All the halogens form covalent hydrides H–X, which are colourless gases at room temperature and which give misty fumes in moist air.

The hydrogen halides are all extremely water soluble. A cubic decimetre of water will dissolve about 270 dm^3 of HCl gas at room temperature, for example. This high solubility is largely due to the high hydration enthalpy of the hydrogen ions and of the relatively small halide ions, which compensates for the bond dissociation energy of the molecule (Table 7.3). HCl, HBr and HI are all strong acids, becoming stronger as the halogen atom gets bigger. The **strength** of an acid refers to the extent to which an acid is dissociated in solution into its ions. Strong acids are nearly completely dissociated; most of their molecules have broken up into ions. Weak acids are not completely dissociated; the weak acid ethanoic acid, CH_3COOH, has only about 2% of its molecules dissociated into its ions in a 0.10 mol dm^{-3} solution. The acidity is a result of the following reaction:

$$HX(aq) + H_2O(l) \rightarrow H_3O^+(aq) + X^-(aq)$$

DEFINITION

The **hydration enthalpy** of an ion is the enthalpy change accompanying the dissolution of one mole of gaseous ions into an aqueous solution sufficiently dilute so that further addition of water causes no further enthalpy change.

and the extent to which this dissociation occurs is largely dependent on the H–X bond strength. The differences in acid strength are not practically evident in water. HI is seldom come across, being rather prone to air oxidation to give iodine so that it does not keep very well. It has no acidic properties to give it advantage over HCl.

Table 7.3 *Bond enthalpies of hydrogen halides*

Bond	H–F	H–Cl	H–Br	H–I
Bond enthalpy/ kJ mol^{-1}	562	431	366	299

BACKGROUND

Hydrofluoric acid is the odd one out; aqueous HF behaves as a weak acid. There is argument as to whether HF(aq) really is weak or whether it simply appears to be so because of other properties not directly related to its acidity. The considerable hydrogen bonding in liquid HF was explored in Chapter 3, and the effect of this on its boiling temperature noted. The H–F bond is indeed the strongest of the four, but this factor alone is probably not enough to explain the difference in acid strength. The ability of fluoride ions to form hydrogen bonds with undissociated HF molecules is also a factor:

$$H\text{–}F(aq) + F^-(aq) \rightarrow [F \cdots H - F]^-(aq)$$

This reaction effectively prevents HF molecules from dissociating. Salts such as KHF_2 can be prepared..

Hydrogen fluoride dissolves glass, and for this reason is popularly thought to be the most corrosive of acids. It gives extremely unpleasant and painful burns which take a long time to heal, but is not especially corrosive otherwise. Its ability to attack glass comes from the formation of the ion SiF_6^{2-}, which is water soluble; other halide ions are too large to fit around the silicon atom, and so do not form similar complexes. Solutions of HF are kept in poly(ethene) bottles

Tests for halide ions

The usual test for halide ions other than fluoride uses the insolubility of silver halides. The test solution is made acidic with dilute nitric acid to ensure the removal of carbonate or sulphite ions which would interfere by giving a spurious precipitate, and then aqueous silver nitrate solution is added. A white precipitate indicates chloride, a pale creamy precipitate bromide, and a yellow precipitate iodide. The general equation representing the reaction is

$$X^-(aq) + Ag^+(aq) \rightleftharpoons AgX(s)$$

BACKGROUND

The precipitates can further be distinguished by their reaction with ammonia solution. The effect of ammonia in this test depends on the equilibrium where ammonia complexes with aqueous silver ions to form diamminesilver(I) :

$$Ag^+(aq) + 2NH_3(aq) \rightleftharpoons [Ag(NH_3)_2]^+(aq)$$

This reaction reduces the concentration of silver ions in the solution. All these silver halides are sparingly soluble; the solubilities are given in Table 7.4.

Table 7.4 ***The solubilities of silver halides at 25°C***

Halide	AgCl	AgBr	AgI
Solubility/mol dm^{-3}	1.34×10^{-5}	8.80×10^{-7}	9.10×10^{-9}

Silver chloride is sufficently soluble to enable the equilibrium given above to remove enough silver ions from the solution using dilute ammonia so that the solubility of silver chloride is not exceeded. Silver bromide is less soluble and so concentrated ammonia is required to move the equilibrium sufficiently to the right to have the same effect. Silver iodide is too insoluble, so the equilibrium cannot remove sufficient silver ions in this case and the precipitate does not dissolve.

These results are summarised in the test box. The test cannot be used for fluorides since silver fluoride is water soluble.

TEST

	chloride	bromide	iodide
Addition of $Ag^+(aq)$	white precipitate	pale yellow precipitate	yellow precipitate
Addition of dilute NH_3	precipitate dissolves	no change	no change
Addition of conc NH_3	precipitate dissolves	precipitate dissolves	no change

An alternative method of testing for bromide and iodide ions is to oxidise them to the respective halogen, then shake with an immiscible organic solvent such as hexane. The test solution is made acidic with nitric acid, and a little hexane added, followed by a little sodium chlorate(I) solution. On shaking it will produce a brown organic layer if bromide ions are present, and a purple one if iodide ions are present.

Reactions of halide salts with concentrated sulphuric acid

The halides react with concentrated sulphuric acid in a manner determined by the ease of oxidation of the halide ion. Thus sodium chloride does not give chlorine since sulphuric acid will not oxidise chloride ions; bromides and iodides do, however, give the halogen.

Sodium chloride reacts with concentrated sulphuric acid at room temperature to give hydrogen chloride as misty fumes, and sodium hydrogen sulphate:

$$NaCl(s) + H_2SO_4(l) \rightarrow HCl(g) + NaHSO_4(s)$$

Both bromide ions and iodide ions are stronger reducing agents than chloride ions; being larger, their electrons are removed more readily. Bromide ions reduce sulphuric acid to sulphur dioxide. The products from the reaction are a mixture of hydrogen bromide and bromine:

$$NaBr(s) + H_2SO_4(l) \rightarrow HBr(g) + NaHSO_4(s)$$

$$2HBr(g) + H_2SO_4(l) \rightarrow Br_2(l) + 2H_2O(l) + SO_2(g)$$

The state symbols should be cautiously interpreted; the reaction mixture is a mess. Some sulphur dioxide will remain in solution as sulphurous acid, and some of the bromine will be lost as a brown gas.

QUESTION

Explain in terms of atomic properties why the reducing ability of halide ions changes $Cl^- < Br^- < I^-$.

Iodide ions are bigger than bromide ions and are even more reducing; they will reduce sulphuric acid in three ways, all of which occur if sodium iodide is added to concentrated sulphuric acid at room temperature. No usable amounts of HI are produced, and the mixture is even more of a mess, being sludgily brown with purple iodine vapour being evolved, together with the delicate overtones of bad eggs from hydrogen sulphide.

BACKGROUND

The four reactions are:

$$NaI(s) + H_2SO_4(l) \rightarrow HI(g) + NaHSO_4(s)$$

$$8HI(g) + H_2SO_4(l) \rightarrow 4I_2(s) + H_2S(g) + 4H_2O(l)$$

$$6HI(g) + H_2SO_4(l) \rightarrow 3I_2(s) + S(s) + 4H_2O(l)$$

$$2HI(g) + H_2SO_4(l) \rightarrow I_2(s) + SO_2(g) + 2H_2O(l)$$

The brown colour is due to the production of the tri-iodide ion from iodine and excess iodide:

$$I_2 + I^- \rightleftharpoons I_3^-$$

Positive oxidation states in the chemistry of chlorine

Chlorine is the third most electronegative element after fluorine and oxygen, so when in combination with these elements it shows positive oxidation states. The commonest compounds show chlorine (+1) in sodium chlorate(I) or sodium hypochlorite NaOCl, and (+5) in sodium chlorate(V), $NaClO_3$, often just called sodium chlorate.

Sodium chlorate(I) is common household bleach, made from chlorine and sodium hydroxide:

$$\underset{}{2NaOH(aq)} + \underset{(0)}{Cl_2(aq)} \rightarrow \underset{(+1)}{NaOCl(aq)} + \underset{(-1)}{NaCl(aq)} + H_2O(l)$$

Ionic equations are often preferable:

$$2OH^-(aq) + \underset{(0)}{Cl_2(aq)} \rightarrow \underset{(+1)}{OCl^-(aq)} + \underset{(-1)}{Cl^-(aq)} + H_2O(l)$$

DEFINITION

Disproportionation occurs when a given type of atom (either as the element or in a compound) undergoes a reaction in which it is both oxidised and reduced.

The oxidation states of the chlorine are shown; the element has been simultaneously oxidised and reduced, so has undergone disproportionation. If the reaction is carried out at a higher temperature, around 80°C, then the reaction occurring is:

$$\underset{}{6NaOH(aq)} + \underset{(0)}{3Cl_2(aq)} \rightarrow \underset{(+5)}{NaClO_3(aq)} + \underset{(-1)}{5NaCl(aq)} + 3H_2O(l)$$

The elemental chlorine has disproportionated in a different manner. The chlorate(I) ion will also disproportionate if heated in aqueous solution:

$$\underset{(+1)}{3OCl^-(aq)} \rightarrow \underset{(-1)}{2Cl^-(aq)} + \underset{(+5)}{ClO_3^-(aq)}$$

The oxyacids themselves, HOCl and $HClO_3$, are seldom met; the oxyanions are oxidising agents in acidic solution, though, and are sometimes used as more convenient oxidising agents in the laboratory than the halogens themselves:

$$ClO^-(aq) + 2H^+(aq) + 2e^- \rightarrow Cl^-(aq) + H_2O(l)$$

$$ClO_3^-(aq) + 6H^+(aq) + 6e^- \rightarrow Cl^-(aq) + 3H_2O(l)$$

Oxidation reactions of the halogens

All of the halogens are strong oxidising agents, the oxidising power decreasing as the size of the halogen increases.

Any halogen will oxidise the ions of those halogens below it in the group. Thus chlorine will oxidise bromide and iodide ions, and bromine will oxidise iodide ions.

Bromine manufacture: bromine is extracted by oxidation with chlorine of the bromide ions present in seawater. About 30,000 tonnes of bromine are produced annually in the UK, mostly for the manufacture of 1,2-dibromoethane which is used as an additive for leaded petrol and as an intermediate in the production of ethane-1,2-diol for antifreeze and hydraulic fluid. The details of the bromine extraction are not important; the reaction is essentially:

$$2Br^-(aq) + Cl_2(aq) \rightarrow Br_2(aq) + 2Cl^-(aq)$$

Questions

1. Draw (small) graphs of the properties in Table 7.1 vs atomic number.
2. Draw (small) graphs of the properties in Table 7.2 vs atomic number.
3. Draw a dot-and-cross diagram to show the bonding in the ion I_3^-, and state its shape, with reasons.
4. Give the equation representing the reaction of fluorine with water at room temperature.

Assessment questions

The following questions are Edexcel AS questions. The earlier AS papers were out of 75; current ones are out of 60, just by reducing the number of questions. The earlier papers have been left as they were set.

1 (*a*) (i) Write the equation for the reaction of lithium with water. **[2]**

(ii) Describe what you would expect to see during the reaction. **[2]**

(*b*) State the number of protons, neutrons and electrons in a $^{7}_{3}Li^{+}$ ion. **[3]**

(*c*) The mass spectrum of lithium shows two peaks. Their mass/charge ratios and percentage abundance are shown below.

Mass/charge	% Abundance
6.02	7.39
7.02	92.61

Calculate the relative atomic mass of lithium, giving your answer to three significant figures. **[2]**

(*d*) Describe a test that you would do to distinguish between solid lithium chloride and solid sodium chloride. Clearly state what you would do and what you would see with both substances. **[3]**

(Total 12 marks)
(January 2003)

2 (*a*) Complete the following table:

Particle	Relative charge	Relative mass
Proton		1
Electron	–1	
Neutron		1

[3]

(*b*) State the number of each of the above particles present in one molecule of CH_4, showing clearly how you arrive at your answer. **[3]**

(*c*) Complete the electronic configuration of a chlorine atom.

$1s^2$ **[1]**

(*d*) Give the **formula** of the chlorine species composed of 17 protons, 20 neutrons and 16 electrons. **[2]**

(*e*) Write one equation in each case to represent the change occurring when the following quantities are measured.

(i) The first electron affinity of sulphur. **[2]**

(ii) The first ionisation energy of sulphur **[1]**

(*f*) Explain why the first ionisation energy of chlorine is higher than that of sulphur. **[2]**

(Total 14 marks)
(May 2002)

3 (*a*) The first ionisation energy of potassium is $+419\ kJ\ mol^{-1}$ and that of sodium is $+496\ kJ\ mol^{-1}$.

(i) Define the term **first ionisation energy**. **[3]**

(ii) Explain why the first ionisation energy of potassium is only a little less than the first ionisation energy of sodium. **[3]**

(*b*) Potassium forms a superoxide, KO_2. This reacts with carbon dioxide according to the equation:

$$4KO_2(s) + 2CO_2(g) \rightarrow 2K_2CO_3(s) + 3O_2(g)$$

Carbon dioxide gas was reacted with 4.56 g of potassium superoxide.

(i) Calculate the amount, in moles, of KO_2 in 4.56 g of potassium superoxide. **[2]**

(ii) Calculate the amount, in moles, of carbon dioxide that would react with 4.56 g of potassium superoxide. **[1]**

(iii) Calculate the volume of carbon dioxide, in dm^3, that would react with 4.56 g of potassium superoxide. Assume that 1.00 mol of a gas occupies 24 dm^3 under the conditions of the experiment. **[1]**

(iv) What volume of oxygen gas, in dm^3, measured under the same conditions of pressure and temperature, would be released? **[1]**

(Total 11 marks)
(January 2003)

4 (*a*) (i) Calculate the number of moles of potassium nitrate, KNO_3, in 10.1 g of KNO_3. **[1]**

(ii) Potassium nitrate, KNO_3, can be prepared from potassium hydroxide solution as shown in the following equation:

$$KOH(aq) + HNO_3(aq) \rightarrow KNO_3(aq) + H_2O(1)$$

Calculate the minimum volume, in cm^3, of 2.00 mol dm^{-3} KOH required to produce 10.1 g of KNO_3. **[2]**

(iii) Potassium nitrate decomposes, when heated, to produce oxygen.

$$2KNO_3(s) \rightarrow 2KNO_2(s) + O_2(g)$$

Calculate the volume of oxygen gas, in dm^3, produced when 10.1 g of potassium nitrate decomposes in this way.

(1 mole of gas has a volume of 24 dm^3 under the conditions of the experiment.) **[2]**

(*b*) A compound of potassium and oxygen contains 70.9% potassium.

(i) Calculate the empirical formula of this compound, using the data above and the periodic table. **[3]**

(ii) 0.200 moles of this compound has a mass of 22.0 g. Use this information to help you deduce the molecular formula of this compound. **[2]**

(Total 10 marks)
(May 2002)

5 Nitrogen and phosphorus are in the same group of the Periodic Table. Phosphorus and hydrogen form the compound phosphine, PH_3, and nitrogen and hydrogen form ammonia, NH_3.

(*a*) (i) State the number of bond pairs and lone pairs of electrons in a molecule of phosphine. **[2]**

(ii) Use your answer to (i) to draw the shape of the molecule and indicate on your diagram the approximate HPH bond angle that you would expect. **[2]**

(*b*) The boiling temperature of ammonia is –33 °C and that of phosphine is –88 °C.

(i) List all the intermolecular forces that exist between molecules of ammonia. **[2]**

(ii) Explain why the boiling temperature of phosphine is lower than that for ammonia. **[2]**

(*c*) Ammonia forms a dative covalent bond with H^+ ions to form the ammonium ion, NH_4^+.

(i) Explain what is meant by the term **dative covalent bond**. **[2]**

(ii) What part of the ammonia molecule enables it to form such a bond? **[1]**

(iii) State and explain the shape of the ammonium ion, NH_4^+. **[3]**

(Total 14 marks)
(January 2003)

6 (*a*) State the structure of, and the type of bonding in, the following substances. Draw labelled diagrams to illustrate your answers.

(i) Graphite **[4]**

(ii) Sodium chloride **[3]**

(*b*) Explain why **both** graphite and sodium chloride have high melting temperatures. **[3]**

(*c*) (i) Explain why graphite is able to conduct electricity in the solid state. **[2]**

(ii) Explain why sodium chloride conducts electricity in the liquid state. **[1]**

(Total 13 marks)
(May 2002)

ASSESSMENT QUESTIONS

7 The table below shows the melting temperatures of the elements of period 3.

	Na	Mg	Al	Si	P	S	Cl	Ar
m.p./°C	98	650	660	1410	44	119	–101	–189

(*a*) Write the structural type for each element shown. **[2]**

(*b*) Explain why the melting temperature of sodium is so much lower than that of magnesium or of aluminium. **[3]**

(*c*) (i) Explain the very low melting temperature of argon. **[1]**

(ii) Phosphorus exists as P_4, sulphur as S_8. Explain the difference in the melting temperature of these substances. **[2]**

(Total 8 marks)
(January 2002)

8 (*a*) Seawater contains aqueous bromide ions. During the manufacture of bromine, seawater is treated with chlorine gas and the following reaction occurs:

$$2Br^- + Cl_2 \rightarrow Br_2 + 2Cl^-$$

(i) Explain the term **oxidation** in terms of electron transfer. **[1]**

(ii) Explain the term **oxidising agent** in terms of electron transfer. **[1]**

(iii) State which of the elements chlorine or bromine is the stronger oxidising agent and explain the importance of this in the extraction of bromine from seawater, as represented in the equation above. **[2]**

(*b*) When sodium chlorate(I), NaClO, is heated, sodium chlorate(V) and sodium chloride are formed.

(i) Write the **ionic** equation for this reaction. **[2]**

(ii) What type of reaction is this? **[1]**

(*c*) During one process for the manufacture of iodine the following reaction occurs:

$$2IO_3^- + 5SO_2 + 4H_2O \rightarrow I_2 + 8H^+ + 5SO_4^{2-}$$

(i) Deduce the oxidation number of sulphur in:

SO_2

SO_4^{2-}

[2]

(ii) Use your answers to part (c)(i) to explain whether SO_2 has been oxidised or reduced in the above reaction. **[1]**

(iii) Name a reagent that could be used to confirm that a solution contains iodine, and state what would be **seen**. **[2]**

(Total 12 marks)
(June 2001)

9 (*a*) (i) State how a flame test would distinguish between samples of calcium nitrate, $Ca(NO_3)_2$ and barium nitrate, $Ba(NO_3)_2$. **[2]**

(ii) Explain the origin of the flame colour. **[3]**

(*b*) Write the equation for the action of heat on barium nitrate. **[2]**

(*c*) (i) What is meant by the term **polarising power** as applied to cations? **[2]**

(ii) Give **two** factors which affect the polarising power of cations. **[2]**

(iii) Use this information to explain why it is easier to decompose magnesium nitrate than barium nitrate by heating. **[3]**

(Total 14 marks)
(January 2002)

10 (*a*) Sodium reacts with cold water.

(i) What would you **see** as the reaction proceeds? **[2]**

(ii) Write the balanced chemical equation for this reaction. **[2]**

(*b*) Calculate the volume of gas produced if 3.0 g of sodium reacts with an excess of water.

(One mole of any gas at the temperature and pressure of the experiment occupies 24 dm^3.) **[3]**

(Total 7 marks)
(January 2002)

11 (*a*) The compounds lithium chloride, sodium bromide and potassium iodide can be distinguished from one another by the use of flame tests.

(i) Give the flame colours for lithium, sodium and potassium. **[3]**

(ii) Explain the origin of the colours in flame tests. **[2]**

(*b*) These compounds can also be distinguished from one another by the use of concentrated sulphuric acid.

(i) State what would be **seen** when concentrated sulphuric acid is added to separate solid samples of each of these compounds.

Lithium chloride;

Sodium bromide;

Potassium iodide;

[4]

(ii) Write an equation, including the state symbols, for the reaction between solid lithium chloride and concentrated sulphuric acid. **[2]**

(Total 11 marks)
(June 2001)

12 (*a*) Bromine is a *p*-block element. Define the term ***p*-block element**. **[1]**

(*b*) (i) Give the colour and physical state of bromine at room temperature. **[2]**

(ii) State what you would see when aqueous bromine is added to a solution of potassium iodide. **[1]**

(*c*) Aqueous bromine will oxidise Fe^{2+} ions to Fe^{3+} ions.

(i) Write the ionic half-equation for the reduction of bromine to bromide ions. **[1]**

(ii) Write the ionic half-equation for the oxidation of Fe^{2+} ions to Fe^{3+} ions. **[1]**

(iii) Hence write the overall ionic equation for the reaction of Fe^{2+} ions with bromine. **[1]**

(*d*) Chlorine and bromine react with aqueous sodium hydroxide in a similar way at room temperature.

(i) Write the equation for the reaction of bromine with aqueous sodium hydroxide. **[2]**

(ii) What **type** of reaction is this? **[1]**

(*e*) Potassium bromide, KBr, reacts with potassium bromate, $KBrO_3$, in the presence of dilute sulphuric acid to form bromine, potassium sulphate and water.

$$5KBr + KBrO_3 + 3H_2SO_4 \rightarrow 3Br_2 + 3K_2SO_4 + 3H_2O$$

(i) Give the oxidation numbers of bromide in

KBr;

$KBrO_3$;

Br_2

[3]

(ii) Which substance in this reaction is the oxidising agent? Give a reason for your choice. **[2]**

(Total 15 marks)
(January 2003)

13 (*a*) Describe a chemical test for chlorine. **[2]**

(*b*) Chlorine gas reacts with hydrogen to form hydrogen chloride gas, HCl.

(i) Name the type of bonding in hydrogen chloride gas. **[1]**

(ii) Explain why hydrogen chloride gas is soluble in water. **[2]**

(*c*) A solution of hydrogen chloride reacts with solid calcium carbonate. Write the equation for this reaction. Include the state symbols. **[2]**

(Total 7 marks)
(January 2003)

ASSESSMENT QUESTIONS

14 (*a*) A compound of sodium, chlorine and oxygen contains, by mass, 21.6% Na, 33.3% Cl and 45.1% O. Show that this is consistent with the formula $NaClO_3$. **[2]**

(*b*) $NaClO_3$ can be obtained from NaOCl(aq) by a disproportionation reaction on heating.

(i) Give the **ionic** equation for this disproportionation reaction. **[2]**

(ii) By a consideration of the oxidation numbers of the **chlorine** in the various species, show why the reaction in (i) is disproportionation. **[4]**

(*c*) Chlorine is used in the extraction of bromine from seawater.

(i) Give the half-equation for the reduction of chlorine. **[1]**

(ii) Give the half-equation for the oxidation that is occurring given that the overall equation for the reaction is:

$$Cl_2(aq) + 2Br^-(aq) \rightarrow Br_2(aq) + 2Cl^-(aq)$$

[1]

(Total 10 marks)
(January 2002)

15 (*a*) Hydrogen chloride can be made from sodium chloride and concentrated sulphuric acid. Write a balanced chemical equation to represent this reaction. **[1]**

(*b*) (i) How would you confirm that a solution said to be HCl(aq) contained chloride ions? **[3]**

(ii) Hydrogen chloride is soluble in water. Explain why the solution is acidic. **[2]**

(*c*) (i) Give a chemical test for chlorine, stating what you would do and what you would see. **[2]**

(ii) Hydrogen chloride can be oxidised to chlorine by lead(IV) oxide, PbO_2. What are the oxidation numbers of lead and of chlorine in the reaction?

$$PbO_2 + 4HCl \rightarrow PbCl_2 + Cl_2 + 2H_2O$$

[2]

(*d*) Sodium iodide reacts with concentrated sulphuric acid to give iodine, not hydrogen iodide. Explain why iodides react differently from chlorides in this case. **[2]**

(Total 12 marks)
(January 2002)

16 (*a*) Define the term **first ionisation energy** for magnesium. **[3]**

(*b*) The logarithm of successive ionisation energies for magnesium is plotted in the graph below.

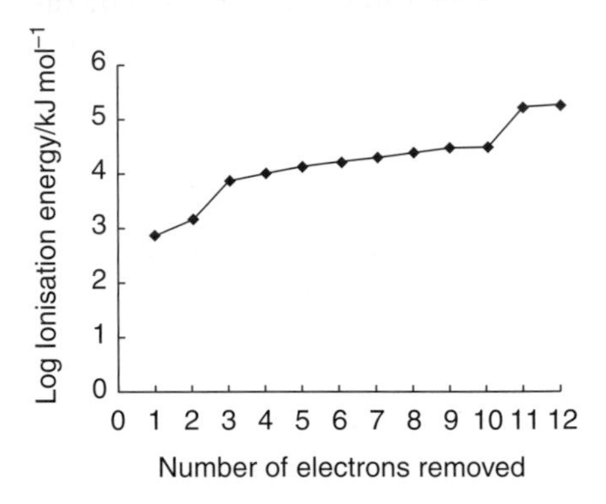

Explain what this graph tells you about the electron arrangement in the magnesium atom. **[3]**

(*c*) (i) Give the full electronic configuration of magnesium using the s,p,d notation. **[1]**

(ii) Explain why all isotopes of magnesium have the **same** chemical properties. **[2]**

(Total 9 marks)
(January 2002)

17 (*a*) (i) What is meant by the **mass number** of an atom? **[1]**

(ii) Define the term **relative atomic mass**. **[2]**

(iii) What are isotopes? **[2]**

(*b*) Magnesium has three isotopes. The mass spectrum of magnesium shows peaks at *m/e* 24 (78.60%), 25 (10.11%), and 26 (11.29%). Calculate the relative atomic mass of magnesium to 4 significant figures. **[2]**

(Total 7 marks)
(January 2002)

18 (*a*) Draw the shape of each of the following molecules and mark on the diagram a value for the bond angle in each case.

(i) CH_4 **[2]**

(ii) NH_3 **[2]**

(iii) $BeCl_2$ **[2]**

(*b*) Name the strongest type of intermolecular force present in separate samples of each of the following substances:

(i) CH_4 **[1]**

(ii) HF **[1]**

(*c*) State and explain which of the substances CH_4 and HF has the higher boiling temperature. **[2]**

(Total 10 marks)
(May 2002)

19 (*a*) Boron forms the chloride BCl_3. Draw a dot-and-cross diagram for BCl_3. **[1]**

(*b*) (i) Draw the shape of the BCl_3 molecule. **[1]**

(ii) Explain why BCl_3 has this shape. **[2]**

(*c*) (i) The B—Cl bond is polar due to the different electronegativity of the atoms. Explain what is meant by the term **electronegativity**. **[2]**

(ii) The B—Cl bond is polar. Explain why BCl_3 is **not** a polar molecule. **[2]**

(Total 8 marks)
(January 2002)

20 (*a*) When the Group 2 element calcium is added to water, calcium hydroxide and hydrogen are produced.

Write an equation for the reaction. **[1]**

(*b*) State the trend in solubility of the hydroxides of the Group 2 elements as the atomic mass of the metal increases. **[1]**

(*c*) (i) Define the term **first ionisation energy**, and write an equation to represent the change occurring when the first ionisation energy of calcium is measured. **[4]**

(ii) State and explain the trend in the first ionisation energy of the Group 2 elements. **[3]**

(Total 9 marks)
(June 2001)

21 (*a*) (i) Complete the electronic configuration of a sulphur atom.

$1s^2$ **[1]**

(ii) Deduce the number of neutrons in the nucleus of an atom of ^{32}S. **[1]**

(*b*) (i) Define the term **first electron affinity**. **[3]**

(ii) The following equation represents the change occurring when the second electron affinity of sulphur is measured.

$$S^-(g) + e^- \rightarrow S^{2-}(g)$$

Explain why the **second** electron affinity of an element is endothermic. **[2]**

(Total 7 marks)
(June 2001)

22 (*a*) When a sample of copper is analysed using a mass spectrometer, its atoms are ionised and then accelerated.

(i) Explain how the atoms of the sample are ionised. **[2]**

(ii) State how the resulting ions are then accelerated. **[1]**

(*b*) For a particular sample of copper two peaks were obtained in the mass spectrum.

Peak at *m/e*	Relative abundance
63	69.1
65	30.9

(i) Give the formula of the species responsible for the peak at $m/e = 65$. **[1]**

(ii) State why **two** peaks, at *m/e* values of 63 and 65, were obtained in the mass spectrum. **[1]**

(iii) Calculate the relative atomic mass of this sample of copper, using the table of results above. **[2]**

(Total 7 marks)
(June 2001)

23 (*a*) Iron has a number of isotopes. One of them has the electronic configuration $[Ar]3d^6 4s^2$, an atomic number of 26 and a mass number of 56.

(i) Which of these pieces of information would be the most use in helping a chemist decide on the likely chemical reactions of iron? **[1]**

(ii) State how many of each of the following particles is found in an atom of ^{56}Fe.

Protons; electrons; neutrons; **[2]**

(iii) What are isotopes? **[2]**

(*b*) The relative atomic mass of a sample of iron may be found by using a mass spectrometer to determine the isotopic composition.

(i) The diagram below represents a low-resolution mass spectrometer in which 4 areas have been identified. For each part of the intrument state what happens there.

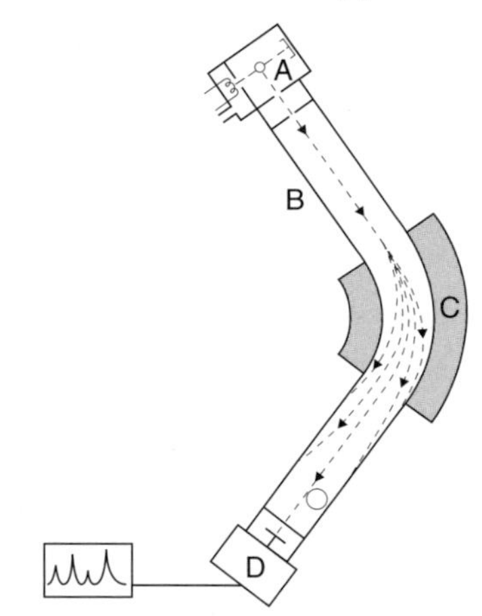

Area A;

Area B;

Area C;

Area D; **[4]**

(ii) In such a determination the following isotopic composition was found.

Isotope	Percentage composition
^{54}Fe	5.8
^{56}Fe	91.6
^{57}Fe	2.2
^{58}Fe	0.33

Calculate the relative atomic mass of this sample of iron, giving your answer to 2 decimal places only. **[2]**

(Total 11 marks)

(January 2001)

24 (*a*) The first ionisation energy of chlorine is +1260 kJ mol^{-1} and the first electron affinity of chlorine is –364 kJ mol^{-1}.

(i) Define the term *first ionisation energy.* **[3]**

(ii) State and explain the general trend in the values of the first ionisation energy for the elements across the period sodium to argon in the Periodic Table. **[2]**

(iii) Write an equation to show the change occurring when the first electron affinity of chlorine is measured. **[2]**

(*b*) 0.5 moles of chlorine were passed into an aqueous solution containing 0.66 moles of EACH of sodium bromide and sodium iodide. Assuming that all the chlorine reacted calculate:

(i) the number of moles of iodine produced;

(ii) the number of moles of bromine produced. **[5]**

(Total 12 marks)

(January 2001)

25 (*a*) Compound **A**, consisting of carbon and hydrogen only, was found to contain 80.0% carbon by mass.

(i) Calculate the empirical formula of compound **A**, using the data above and the Periodic Table. **[3]**

(ii) The relative molecular mass of compound **A** was found to be 30. Use this information to deduce the molecular formula of compound **A**. **[1]**

(*b*) Propane has the molecular formula C_3H_8. Propane burns completely in oxygen to form carbon dioxide and water as shown in the equation.

$$C_3H_8(g) + 5O_2(g) \rightarrow 3CO_2(g) + 4H_2O(g)$$

(i) Calculate the mass of water produced when 110 g of propane burns completely in oxygen. **[3]**

(ii) Calculate the volume of oxygen required to completely burn 110 g of propane. (1 mole of gas has a volume of 24 dm^3 under the conditions of the experiment.) **[2]**

(Total 9 marks)
(June 2001)

26 (*a*) Explain the following observations. Include details of the **bonding** in and the **structure** of each substance.

(i) The melting temperature of diamond is much higher than that of iodine. **[5]**

(ii) Sodium chloride has a high melting temperature (approximately 800 °C). **[3]**

(*b*) Explain why aluminium metal is a good conductor of electricity. **[3]**

(Total 11 marks)
(June 2001)

27 Deduce and draw the shapes of the following molecules or ions. Suggest a value for the bond angle in each case. Give a brief explanation of why each has the shape you give.

(*a*) $\mathbf{SF_6}$ **[3]**

(*b*) $\mathbf{PH_3}$ **[3]**

(*c*) $\mathbf{PF_4^+}$ **[3]**

(Total 9 marks)
(June 2001)

28 (*a*) Consider the following shapes

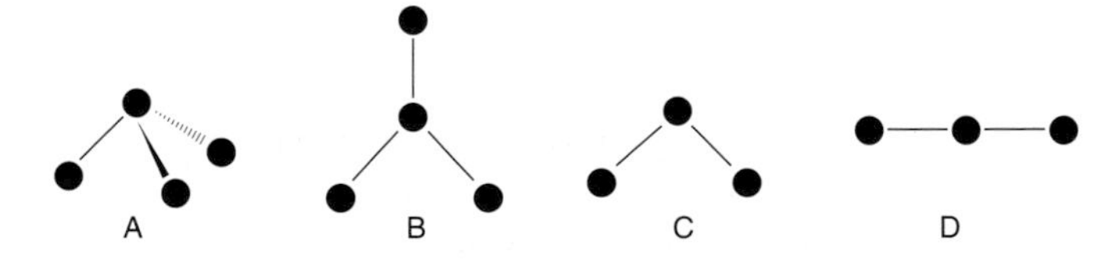

Indicate by a letter (**A, B, C** or **D**) the shape of the following ions or molecules

(i) H_2O

(ii) NH_3

(iii) CO_2

(iv) CO_3^{2-} **[4]**

(*b*) Write balanced equation for the following reactions. You should omit state symbols.

(i) The reaction of potassium metal with water. **[1]**

(ii) The reaction of calcium metal with oxygen. **[1]**

(iii) The reaction of magnesium oxide with nitric acid. **[2]**

(Total 8 marks)
(January 2001)

29 (*a*) Copy and complete the table below.

Element	**Chlorine**	**Bromine**	**Iodine**
State at room temperature			solid
Colour			grey
What would be seen on adding to an aqueous solution of potassium iodide			

[6]

(*b*) Solid iodine has a simple covalent molecular structure.

(i) Define the term covalent bond. **[2]**

(ii) Explain how the covalent structure of iodine leads to it having a relatively low melting temperature. **[3]**

(*c*) The diagram below shows a plot of boiling temperature against relative molecular mass for four hydrogen halides, HF, HCl, HBr, and HI.

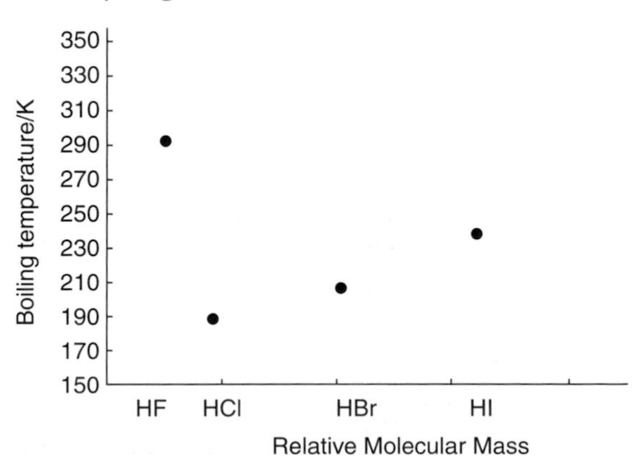

(i) Explain the increase in the boiling temperature from hydrogen chloride, HCl, to hydrogen iodide, HI. **[2]**

(ii) Explain why the boiling temperature of hydrogen fluoride, HF, is higher than the boiling temperature of hydrogen chloride, HCl. **[2]**

(Total 15 marks)
(January 2001)

30 (*a*) (i) Write the equation for the reaction of magnesium metal with chlorine, showing state symbols. **[2]**

(ii) The product of this reaction is ionic. Use this information to explain why it has a relatively high melting temperature. (714 °C) **[2]**

(*b*) Why is magnesium iodide more covalent than magnesium chloride? **[2]**

(*c*) Describe the bonding in magnesium metal. **[3]**

(Total 9 marks)
(January 2001)

31 A solution of a weak acid H_2X was made by dissolving 2.25 g of solid H_2X in water to give 500 cm^3 of solution. On titration, 25.0 cm^3 of this solution was completely neutralised by 25.0 cm^3 of sodium hydroxide solution containing 0.100 mol dm^{-3}.

(*a*) Write an equation for the reaction **[1]**

(*b*) (i) Calculate the number of moles of NaOH in 25.0 cm^3 of 0.100 mol dm^{-3} solution.

(ii) How many moles of H_2X would be required to react with this quantity of NaOH?

(iii) Calculate the relative molecular mass of H_2X.

(iv) A hydrated form of the acid also exists, $H_2X.yH_2O$. A solution containing 6.30 g dm^{-3} of the hydrated acid has the same (molar) concentration as the solution of the anhydrous acid, H_2X, originally used. Using the information and your answer from (*b*), calculate the value of y. **[10]**

(*c*) The presence of sulphur dioxide in the atmosphere is the main cause of acid rain. Outline a method which could be used to estimate quantitatively the concentration of sulphur dioxide in a sample of air. **[5]**

(Total 16 marks)
(June 1996)

32 (*a*) Explain what is meant by the terms **ionic bond**, **covalent bond**, **dative covalent bond**.

Ionic bond; **[2]**

Covalent bond; **[2]**

Dative covalent bond. **[2]**

(*b*) Indicate whether each of the following molecules has an overall polarity, giving your reasons. **[6]**

	Yes/No
Tetrachloromethane, CCl_4	
Methane, CH_4	
Ethanol, CH_3CH_2OH	

(*c*) The atomic and ionic radii for a number of elements are given below.

Atom	*Radius/nm*	*Ion*	*Radius/nm*
Na	0.157	Na^+	0.102
Mg	0.136	Mg^{2+}	0.072
Al	0.125	Al^{3+}	0.053
F	0.071	F^-	0.133
Cl	0.099	Cl^-	0.180
I	0.133	I^-	0.216

Suggest an explanation for each of the following:

(i) The magnesium atom is smaller than the sodium atom. **[3]**

(ii) The sodium ion is smaller than the sodium atom. **[3]**

(iii) Aluminium fluoride is ionic, aluminium iodide is covalent. **[5]**

(Total 23 marks)
(January 1996)

33 (*a*) Complete the following table. **[5]**

	Na	Mg	Al	Si	P	S	Cl
Formula of chloride							
Bonding in each chloride							

(*b*) A chloride of phosphorus, **A**, contains 22.5% **P** by mass. Reaction of this compound with more chlorine gives another chloride, **B**, containing 14.9% **P**.

(i) Calculate the empirical formulae of the two chlorides. **[4]**

(ii) Write an equation for the reaction of **A** with chlorine to give **B**. **[2]**

(*c*) Metal nitrates decompose on heating.

(i) What is the trend in the thermal stability of nitrates of Group 2 metals?
Write the equation for any such decomposition. **[3]**

(ii) Nitrates of Group 1 metals decompose in a different way to those in Group 2. Sodium nitrate on strong heating gives a pale yellow solid, **E**, and a colourless gas which relights a glowing splint. 7.23 g of sodium nitrate gave, in such an experiment, 5.87 g of E and 1.02 dm^3 of gas. This gas contains no nitrogen. Deduce the equation for the thermal decomposition of sodium nitrate. (One mole of gas occupies 24 dm^3 at the temperature and pressure of this experiment.) **[8]**

(Total 22 marks)

(January 1997)

The Periodic Table of Elements

Key

Atomic number
Symbol
Name
Molar mass in g mol^{-1}

Period	1	2						Group					3	4	5	6	7	0
1	1 H Hydrogen 1																	2 He Helium 4
2	3 Li Lithium 7	4 Be Beryllium 9											5 B Boron 11	6 C Carbon 12	7 N Nitrogen 14	8 O Oxygen 16	9 F Fluorine 19	10 Ne Neon 20
3	11 Na Sodium 23	12 Mg Magnesium 24											13 Al Aluminium 27	14 Si Silicon 28	15 P Phosphorus 31	16 S Sulphur 32	17 Cl Chlorine 35.5	18 Ar Argon 40
4	19 K Potassium 39	20 Ca Calcium 40	21 Sc Scandium 45	22 Ti Titanium 48	23 V Vanadium 51	24 Cr Chromium 52	25 Mn Manganese 55	26 Fe Iron 56	27 Co Cobalt 59	28 Ni Nickel 59	29 Cu Copper 63.5	30 Zn Zinc 65.4	31 Ga Gallium 70.4	32 Ge Germanium 73	33 As Arsenic 75	34 Se Selenium 79	35 Br Bromine 80	36 Kr Krypton 84
5	37 Rb Rubidium 85	38 Sr Strontium 88	39 Y Yttrium 89	40 Zr Zirconium 91	41 Nb Niobium 93	42 Mo Molybdenum 96	43 Tc Technetium (99)	44 Ru Ruthenium 101	45 Rh Rhodium 103	46 Pd Palladium 106	47 Ag Silver 108	48 Cd Cadmium 112	49 In Indium 115	50 Sn Tin 119	51 Sb Antimony 122	52 Te Tellurium 128	53 I Iodine 127	54 Xe Xenon 131
6	55 Cs Caesium 133	56 Ba Barium 137	57 ▶ La Lanthanum 139	72 Hf Hafnium 178	73 Ta Tantalum 181	74 W Tungsten 184	75 Re Rhenium 186	76 Os Osmium 190	77 Ir Iridium 192	78 Pt Platinum 195	79 Au Gold 197	80 Hg Mercury 201	81 Tl Thallium 204	82 Pb Lead 207	83 Bi Bismuth 209	84 Po Polonium (210)	85 At Astatine (210)	86 Rn Radon (222)
7	87 Fr Francium (223)	88 Ra Radium (226)	89 ▶▶ Ac Actinium (227)	104 Rf Rutherfordium (261)	105 Db Dubnium (262)	106 Sg Seaborgium (263)	107 Bh Bohrium (264)	108 Hs Hassium (269)	109 Mt Meitnerium (268)	110 Uun Ununnilium (269)	111 Uuu Unununium (272)	112 Uub Ununbium (277)						

▶ Lanthanide elements														
58 Ce Cerium 140	59 Pr Praseodymium 141	60 Nd Neodymium 144	61 Pm Promethium (147)	62 Sm Samarium 150	63 Eu Europium 152	64 Gd Gadolinium 157	65 Tb Terbium 159	66 Dy Dysprosium 163	67 Ho Holmium 165	68 Er Erbium 167	69 Tm Thulium 169	70 Yb Ytterbium 173	71 Lu Lutetium 175	

▶▶ Actinide elements														
90 Th Thorium 232	91 Pa Protactinium (231)	92 U Uranium 238	93 Np Neptunium (237)	94 Pu Plutonium (242)	95 Am Americium (243)	96 Cm Curium (247)	97 Bk Berkelium (245)	98 Cf Californium (251)	99 Es Einsteinium (254)	100 Fm Fermium (253)	101 Md Mendelevium (256)	102 No Nobelium (254)	103 Lr Lawrencium (257)	

Index

Page references in *italics* refer to a table or an illustration.

INDEX